Big Data Analytics

Successfully navigating the data-driven economy presupposes a certain understanding of the technologies and methods to gain insights from Big Data. This book aims to help data science practitioners to successfully manage the transition to Big Data.

Building on familiar content from applied econometrics and business analytics, this book introduces the reader to the basic concepts of Big Data Analytics. The focus of the book is on how to productively apply econometric and machine learning techniques with large, complex data sets, as well as on all the steps involved before analysing the data (data storage, data import, data preparation). The book combines conceptual and theoretical material with the practical application of the concepts using R and SQL. The reader will thus acquire the skills to analyse large data sets, both locally and in the cloud. Various code examples and tutorials, focused on empirical economic and business research, illustrate practical techniques to handle and analyse Big Data.

Key Features:

- Includes many code examples in R and SQL, with R/SQL scripts freely provided online.
- Extensive use of real datasets from empirical economic research and business analytics, with data files freely provided online.
- Leads students and practitioners to think critically about where the bottlenecks are in practical data analysis tasks with large data sets, and how to address them.

The book is a valuable resource for data science practitioners, graduate students and researchers who aim to gain insights from big data in the context of research questions in business, economics, and the social sciences.

Ulrich Matter is an Assistant Professor of Economics at the University of St.Gallen. His primary research interests lie at the intersection of data science, political economics, and media economics. His teaching activities cover topics in data science, applied econometrics, and data analytics. Before joining the University of St. Gallen, he was a Visiting Researcher at the Berkman Klein Center for Internet & Society at Harvard University and a postdoctoral researcher and lecturer at the Faculty for Business and Economics, University of Basel.

CHAPMAN & HALL/CRC DATA SCIENCE SERIES

Reflecting the interdisciplinary nature of the field, this book series brings together researchers, practitioners, and instructors from statistics, computer science, machine learning, and analytics. The series will publish cutting-edge research, industry applications, and textbooks in data science.

The inclusion of concrete examples, applications, and methods is highly encouraged. The scope of the series includes titles in the areas of machine learning, pattern recognition, predictive analytics, business analytics, Big Data, visualization, programming, software, learning analytics, data wrangling, interactive graphics, and reproducible research.

Published Titles

Tree-Based Methods
A Practical Introduction with Applications in R
Brandon M. Greenwell

Urban Informatics
Using Big Data to Understand and Serve Communities
Daniel T. O'Brien

Introduction to Environmental Data Science
Jerry Douglas Davis

Hands-On Data Science for Librarians
Sarah Lin and Dorris Scott

Geographic Data Science with R
Visualizing and Analyzing Environmental Change
Michael C. Wimberly

Practitioner's Guide to Data Science
Hui Lin and Ming Li

Data Science and Analytics Strategy
An Emergent Design Approach
Kailash Awati and Alexander Scriven

Telling Stories with Data
With Applications in R
Rohan Alexander

Data Science for Sensory and Consumer Scientists
Thierry Worch, Julien Delarue, Vanessa Rios De Souza and John Ennis

Big Data Analytics
A Guide to Data Science Practitioners Making the Transition to Big Data
Ulrich Matter

For more information about this series, please visit: https://www.routledge.com/Chapman--HallCRC-Data-Science-Series/book-series/CHDSS

Big Data Analytics
A Guide to Data Science Practitioners
Making the Transition to Big Data

Ulrich Matter

CRC Press is an imprint of the
Taylor & Francis Group, an **informa** business

A CHAPMAN & HALL BOOK

First edition published 2024
by CRC Press
6000 Broken Sound Parkway NW, Suite 300, Boca Raton, FL 33487-2742

and by CRC Press
4 Park Square, Milton Park, Abingdon, Oxon, OX14 4RN

CRC Press is an imprint of Taylor & Francis Group, LLC

Library of Congress Cataloging-in-Publication Data

Names: Matter, Ulrich, author.
Title: Big data analytics : a guide to data science practitioners making the transition to big data / Ulrich Matter.
Description: First edition. | Boca Raton, FL : CRC Press, 2024. | Series: Chapman & Hall/CRC data science series | Includes bibliographical references and index.
Identifiers: LCCN 2023008762 (print) | LCCN 2023008763 (ebook) | ISBN 9781032457550 (hbk) | ISBN 9781032458144 (pbk) | ISBN 9781003378822 (ebk)
Subjects: LCSH: Big data. | Machine learning. | Business--Data processing.
Classification: LCC QA76.9.B45 M3739 2024 (print) | LCC QA76.9.B45 (ebook) | DDC 005.7--dc23/eng/20230519
LC record available at https://lccn.loc.gov/2023008762
LC ebook record available at https://lccn.loc.gov/2023008763

ISBN: 978-1-032-45755-0 (hbk)
ISBN: 978-1-032-45814-4 (pbk)
ISBN: 978-1-003-37882-2 (ebk)

DOI: 10.1201/9781003378822

Typeset in Alegreya Regular font
by KnowledgeWorks Global Ltd.

To Mara. May your unyielding spirit and steadfast determination guide you and show you that with patience and dedication, even complex problems yield to solutions. Here's to you, my little dynamo.

Contents

Preface

Background and goals of this book

In the past ten years, "Big Data" has been frequently referred to as the new "most valuable" resource in highly developed economies, spurring the creation of new goods and services across a range of sectors. Extracting knowledge from large datasets is increasingly seen as a strategic asset for firms, governments, and NGOs. In a similar vein, the increasing size of datasets in empirical economic research (both in number of observations and number of variables) offers new opportunities and poses new challenges for economists and business leaders. To meet these challenges, universities started adapting their curricula in traditional fields such as economics, computer science, and statistics, as well as starting to offer new degrees in data analytics, data science, and data engineering.

However, in practice (both in academia and industry), there is frequently a gap between the assembled knowledge of how to formulate the relevant hypotheses and devise the appropriate empirical strategy (the data analytics side) on one hand and the collection and handling of large amounts of data to test these hypotheses, on the other (the data engineering side). While large, specialized organizations like Google and Amazon can afford to hire entire teams of specialists on either side, as well as the crucially important liaisons between such teams, many small businesses and academic research teams simply cannot. *This is where this book comes into play.*

The primary goal of this book is to help practitioners of data analytics and data science apply their skills in a Big Data setting. By bridging the knowledge gap between the data engineering and analytics sides, this book discusses tools and techniques to allow data analytics and data science practitioners in academia and industry to efficiently handle and analyze large amounts of data in their daily analytics work. In addition, the book aims to give decision makers in data teams and liaisons between analytics teams and engineers a practical overview of helpful approaches to work on Big Data projects. Thus, for the data analytics and data science practitioner in academia or industry, this book can well serve as an introduction and handbook to practical issues of Big Data Analytics. Moreover, many parts of this book originated from lecture materials and interactions with students in my Big Data Analytics course for graduate students in economics at the University of St. Gallen and the University of Lucerne. As such, this book, while not appearing in a classical textbook format, can well serve as a textbook in graduate courses on Big Data Analytics in various degree programs.

A moving target

Big Data Analytics is considered a moving target due to the ever-increasing amounts of data being generated and the rapid developments in software tools and hardware devices used to analyze large datasets. For example, with the recent advent of the Internet of Things (IoT) and the ever-growing number of connected devices, more data is being generated than ever before. This data is constantly changing and evolving, making it difficult to keep up with the latest developments. Additionally, the software tools used to analyze large datasets are constantly being updated and improved, making them more powerful and efficient. As a result, practical Big Data Analytics is a constantly evolving field that requires constant monitoring and updating in order to remain competitive. You might thus be concerned that a couple of months after reading this book, the techniques learned here might be already outdated.

So how can we deal with this situation? Some might suggest that the key is to stay informed of the latest developments in the field, such as the new algorithms, languages, and tools that are being developed. Or, they might suggest that what is important is to stay up to date on the latest trends in the industry, such as the use of large language models (LLMs), as these technologies are becoming increasingly important in the field. In this book, I take a complementary approach. Inspired by the transferability of basic economics, I approach Big Data Analytics by focusing on transferable knowledge and skills. This approach rests on two pillars:

1. First, the emphasis is on investing in a reasonable selection of software tools and solutions that can assist in making the most of the data being collected and analyzed, both now and in the future. This is reflected in the selection of R (R Core Team, 2021) and SQL as the primary languages in this book. While R is clearly one of the most widely used languages in applied econometrics, business analytics, and many domains of data science at the time of writing this book (and this may change in the future), I am confident that learning R (and the related R packages) in the Big Data context will be a highly transferable skill in the long run. I believe this for two primary reasons: a) Recent years have shown that more specialized lower-level software for Big Data Analytics increasingly includes easy-to-use high-level interfaces to R (the packages `arrow` and `sparklyr` discussed in this book are good examples for this development); b) even if

the R-packages (or R itself) discussed in this book will be outdated in a few years, the way R is used as a high-level scripting language (connected to lower-level software and cloud tools) will likely remain in a similar form for many years to come. That is, this book does not simply suggest which current R-package you should use to solve a given problem with a large dataset. Instead, it gives you an idea of what the underlying problem is all about, why a specific R package or underlying specialized software like Spark might be useful (and how it conceptually works), and how the corresponding package and problem are related to the available computing resources. After reading this book, you will be well equipped to address the same computational problems discussed in this book with another language than R (such as Julia or Python) as your primary analytics tool.

2. Second, when dealing with large datasets, the emphasis should be on various Big Data approaches, including a basic understanding of the relevant hardware components (computing resources). Understanding why a task takes so long to compute is not always (only) a matter of which software tool you are using. If you understand why a task is difficult to perform from a hardware standpoint, you will be able to transfer the techniques introduced in this book's R context to other computing environments relatively easily.

The structure of the book discussed in the next subsection is aimed at strengthening these two pillars.

Content and organization of the book

Overall, this book introduces the reader to the fundamental concepts of Big Data Analytics to gain insights from large datasets. Thereby, the book's emphasis is on the practical application of econometrics and business analytics, given large datasets, as well as all of the steps involved before actually analyzing data (data storage, data import, data preparation). The book combines theoretical and conceptual material with practical applications of the concepts using R and SQL. As a result, the reader will gain the fundamental knowledge required to analyze large datasets both locally and in the cloud.

The practical problems associated with analyzing Big Data, as well as the corresponding approaches to solving these problems, are generally presented in the context of applied econometrics and business analytics settings throughout this book. This means that I tend to concentrate on observational data, which is common in economics and business/management research. Furthermore, in terms of statistics/analytics techniques, this context necessitates a special emphasis on regression analysis, as this is the most commonly used statistical tool in applied econometrics. Finally, the context determines the scope of the examples and tutorials. Typically, the goal of a data science project in applied econometrics and business analytics is not to deploy a machine learning model as part of an operational App or web application (as is often the case for many working in data science). Instead, the goal of such projects is to gain insights into a specific economic/business/management question in order to facilitate data-driven decisions or policy recommendations. As a result, the output of such projects (as well as the tutorials/examples in this book) is a set of statistics summarizing the quantitative insights in a way that could be displayed in a seminar/business presentation or an academic paper/business report. Finally, the context will influence how code examples and tutorials are structured. The code examples are typically used as part of an interactive session or in the creation of short analytics scripts (and not the development of larger applications).

The book is organized in four main parts. The first part introduces the reader to the topic of Big Data Analytics from the perspective of a practitioner in empirical economics and business research. It covers the differences between *Big P* and *Big N* problems and shows avenues of how to practically address either.

The second part focuses on the tools and platforms to work with Big Data. This part begins by introducing a set of software tools that will be used extensively throughout the book: (advanced) R and SQL. It then discusses the conceptual foundations of modern computing environments and how different hardware components matter in practical local Big Data Analytics, as well as how virtual servers in the cloud help to scale up and scale out analyses when local hardware lacks sufficient computing resources.

The third part of this book expands on the first components of a data pipeline: data collection and storage, data import/ingestion, data cleaning/transformation, data aggregation, and exploratory data visualization (with a particular focus on Geographic Information Systems, GIS). The chapters in this part of the book discuss fundamental concepts such as the split-apply-combine approach and demonstrate how to use these concepts in practice when working with large datasets in R. Many tutorials and code examples demonstrate how a specific task can be implemented locally as well as in the cloud using comparatively simple tools.

Finally, the fourth part of the book covers a wide range of topics in modern applied econometrics in the context of Big Data, from simple regression estimation and machine learning with Graphics Processing Units (GPUs) to running machine learning pipelines and large-scale text analyses on a Spark cluster.

Prerequisites and requirements

This book focuses heavily on R programming. The reader should be familiar with R and fundamental programming concepts such as loops, control statements, and functions (Appendix B provides additional material on specific R topics that are particularly relevant in this book). Furthermore, the book assumes some knowledge of undergraduate and basic graduate statistics/econometrics. R for Data Science by Wickham and Grolemund (Wickham and Grolemund (2016); this is what our undergraduate students use before taking my Big Data Analytics class), Mostly Harmless Econometrics by Angrist and Pischke (Angrist and Pischke (2008)), and Introduction to Econometrics by Stock and Watson (Stock and Watson (2003)) are all good prep books. Regarding hardware and software requirements, you will generally get along just fine with an up-to-date R and RStudio installation. However, given the nature of this book's topics, some code examples and tutorials might require you to install additional software on your computer. In most of these cases, this additional software is made to work on either Linux, Mac, or Windows machines. In some cases, though, I will point out that certain dependencies might not work on a Windows machine. Generally, this book has been written on a Pop-OS/Ubuntu Linux (version 22.04) machine with R version 4.2.0 (or later) and RStudio 2022.07.2 (or later). All examples (except for the GPU-based computing) have also been successfully tested on a MacBook running on macOS 12.4 and the same R and RStudio versions as above.

Code examples, data sets, and additional documentation

This book comes with several freely available online material. All of which is provided on the book's GitHub repository: https://github.com/umatter/bigdata. The README-file in the repository keeps an up-to-date list with links to R-scripts containing the code examples shown in this book, to data sources and datasets used in this book, as well as to additional files with instructions of how to install some of the packages/software used in this book.

If you are interested in using this book as a text book in one of your courses, you might want to have a look at the GitHub repository hosting my own teaching material, including slides and additional code examples: https://github.com/umatter/bigdata-lecture. All of these materials are published under a CC BY-SA 2.0[1]

---

[1]https://creativecommons.org/licenses/by-sa/2.0/

license. When using these materials, please take note of the corresponding terms: https://creativecommons.org/licenses/by-sa/2.0/.

Thanks

Many thanks go to the students in my past Big Data Analytics classes. Their interest and engagement with the topic, as well as their many great analytics projects, were an important source of motivation to start this book project. I'd also like to thank Lara Spieker, Statistics and Data Science Editor at Chapman & Hall, who was very supportive of this project right from the start, for her encouragement and advice throughout the writing process. I am also grateful to Chris Cartwright, the external editor, for his thorough assistance during the book's drafting stage. Finally, I would like to thank Mara, Marc, and Irene for their love, patience, and company throughout this journey. This book would not have been possible without their encouragement and support in challenging times.

Part I

Setting the Scene: Analyzing Big Data

Introduction

This part of the book introduces you to the topic of Big Data analysis from a variety of perspectives. The goal of this part is to highlight the various aspects of modern econometrics involved in Big Data Analytics, as well as to clarify the approach and perspective taken in this book. In the first step, we must consider what makes data *big*. As a result, we make a fundamental distinction between data analysis problems that can arise from many observations (rows; *big N*) and the problems that can arise from many variables (columns; *big P*).

In a second step, this part provides an overview of the four distinct approaches to Big Data Analytics that are most important for the perspective on Big Data taken in this book: a) statistics/econometrics techniques specifically designed to handle Big Data, b) writing more efficient R code, c) more efficiently using available local computing resources, and d) scaling up and scaling out with cloud computing resources. All of these approaches will be discussed further in the book, and it will be useful to remember the most important conceptual basics underlying these approaches from the overview presented here.

Finally, this section of the book provides two extensive examples of what problems related to (too) many observations or (too) many variables can mean for practical data analysis, as well as how some of the four approaches (a-d) can help in resolving these problems.

1

What is Big in "Big Data"?

In this book, we will think of Big Data as data that is (a) difficult to handle and (b) hard to get value from due to its size and complexity. The handling of Big Data is difficult as the data is often gathered from unorthodox sources, providing poorly structured data (e.g., raw text, web pages, images, etc.) as well as because of the infrastructure needed to store and load/process large amounts of data. Then, the issue of statistical computation itself becomes a challenge. Taken together, getting value/insights from Big Data is related to three distinct properties that render its analysis difficult:

- Handling the *complexity and variety* of sources, structures, and formats of data for analytics purposes is becoming increasingly challenging in the context of empirical economic research and business analytics. On the one hand the ongoing digitization of information and processes boosts the generation and storage of digital data for all kinds of economic and social activity, making such data basically more available for analysis. On the other hand, however, the first order focus of such digitization is typically an end user who directly interacts with the information and is part of these processes, and not the data scientist or data analyst who might be interested in analyzing such data later on. Therefore, the interfaces for systematically collecting such data for analytics purposes are typically not optimal. Moreover, data might come in semi-structured formats such as webpages (i.e., the HyperText Markup Language (HTML)), raw text, or even images – each of which needs a different approach for importing/loading and pre-processing. Anyone who has worked on data analytics projects that build on various types of raw data from various sources knows that a large part of the practical data work deals with how to handle the complexity and variety to get to a useful analytic dataset.

- The *big P* problem: A dataset has close to or even more variables (columns) than observations, which renders the search for a good predictive model with traditional econometric techniques difficult or elusive. For example, suppose you run an e-commerce business that sells hundreds of thousands of products to tens of thousands of customers. You want to figure out from which product category a customer is most likely to buy an item, based on their previous product page visits. That is, you want to (in simple terms) regress an indicator of purchasing from a specific category on indicators for previous product page visits. Given this setup, you would potentially end up with hundreds of thousands of explanatory indicator variables (and potentially even linear combinations of those), while you "only" have tens of thousands of observations (one per user/customer and visit) to estimate your model. These sorts of problems are at the core of the domain of modern predictive econometrics, which shows how machine learning approaches like the lasso estimator can be applied to get reasonable estimates from such a predictive model.

• The *big N* problem: a dataset has massive numbers of observations (rows) such that it cannot be handled with standard data analytics techniques and/or on a standard desktop computer. For example, suppose you want to segment your e-commerce customers based on the traces they leave on your website's server. Specifically, you plan to use the server log files (when does a customer visit the site, from where, etc.) in combination with purchase records and written product reviews by users. You focus on 50 variables that you measure on a daily basis over five years for all 50,000 users. The resulting dataset has $50,000 \times 365 \times 5 = 91,250,000$ rows, with 50 variables (at least 50 columns) – over 4.5 billion cells. Such a dataset can easily take up dozens of gigabytes on the hard disk. Hence it will either not fit into the memory of a standard computer to begin with (import fails), or the standard programs to process and analyze the data will likely be very inefficient and take ages to finish when used on such a large dataset. There are both econometric techniques as well as various specialized software and hardware tools to handle such a situation.

After having a close look at the practical data analytics challenges behind both *big P* and *big N* in Chapter 3, most of this book focuses on practical challenges and solutions related to *big N* problems. However, several of the chapters contain code examples that are primarily discussed as a solution to a *big N* problem, but are shown in the context of econometric/machine learning techniques that are broadly used, for example, to find good predictive models (based on many variables, i.e., *big P*). At the same time, many of the topics discussed in this book are in one way or another related to the difficulties of handling various types of structured, semi-structured, and unstructured data. Hence you will get familiar with practical techniques to deal with *complexity and variety* of data as a byproduct.

2

Approaches to Analyzing Big Data

Throughout the book, we consider four approaches to how to solve challenges related to analyzing big N and big P data. Those approaches should not be understood as mutually exclusive categories; rather they should help us to look at a specific problem from different angles in order to find the most efficient tool/approach to proceed. Figure 2.1 presents an illustrative overview of the four approaches.

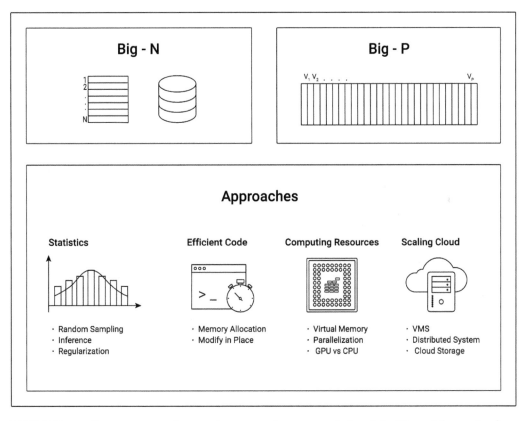

FIGURE 2.1: Four approaches to/perspectives on solving big N problems in data analytics.

1. *Statistics/econometrics and machine learning*: During the initial hype surrounding Big Data/Data Science about a decade ago, statisticians prominently (and justifiably) pointed out that statistics techniques that have always been very useful tools when analyzing "all the data" (the entire

population) is too costly.[1] In simple terms, when confronted with the challenge of answering an empirical question based on a *big N* dataset (which is too large to process on a normal computer), one might ask "why not simply take a random sample?" In some situations this might actually be a very reasonable question, and we should be sure to have a good answer for it before we rent a cluster computer with specialized software for distributed computing. After all, statistical inference is there to help us answer empirical questions in situations where collecting data on the entire population would be practically impossible or simply way too costly. In today's world, digital data is abundant in many domains, and the collection is not so much the problem any longer; but our standard data analytics tools are not made to analyze such amounts of data. Depending on the question and data at hand, it might thus make sense to simply use well-established "traditional" statistics/econometrics in order to properly address the empirical question. Note, though, that there are also various situations in which this would not work well. For example, consider online advertising. If you want to figure out which user characteristics make a user significantly more likely to click on a specific type of ad, you likely need hundreds of millions of data points because the expected probability that a specific user clicks on an ad is generally very low. That is, in many practical Big Data Analytics settings, you might expect rather small effects. Consequently, you need to rely on a big N dataset in order to get the statistical power to distinguish an actual effect from a zero effect. However, even then, it might make sense to first look at newer statistical procedures that are specifically made for big N data before renting a cluster computer. Similarly, traditional statistical/econometric approaches might help to deal with big P data, but they are usually rather inefficient or have rather problematic statistical properties in such situations. However, there are also well-established machine learning approaches to better address these problems. In sum, before focusing on specialized software like Apache Hadoop or Apache Spark and scaling up hardware resources, make sure to use the adequate statistical tools for a Big Data situation. This can save a lot of time and money. Once you have found the most efficient statistical procedure for the problem at hand, you can focus on how to compute it.

2. *Writing efficient code*: No matter how suitable a statistical procedure is theoretically to analyze a large dataset, there are always various ways to implement this procedure in software. Some ways will be less efficient than others. When working with small or moderately sized datasets, you

---

[1] David Donoho has nicely summarized this critique in a paper titled "50 Years of Data Science" (Donoho (2017)), which I warmly recommend.

might not even notice whether your data analytics script is written in an efficient way. However, it might get uncomfortable to run your script once you confront it with a large dataset. Hence the question you should ask yourself when taking this perspective is, "Can I write this script in a different way to make it faster (but achieve the same result)?" Before introducing you to specialized R packages to work with large datasets, we thus look at a few important aspects of how to write efficient/fast code in R.

3. *Using limited local computing resources more efficiently*: There are several strategies to use the available local computing resources (your PC) more efficiently, and many of those have been around for a while. In simple terms, these strategies are based on the idea of more explicitly telling the computer how to allocate and use the available hardware resources as part of a data analytics task (something that is usually automatically taken care of by the PC's operating system). We will touch upon several of these strategies – such as multi-core processing and the efficient use of virtual memory – and then practically implement these strategies with the help of specialized R packages. Unlike writing more efficient R code, these packages/strategies usually come with an overhead. That is, they help you save time only after a certain threshold. In other words, not using these approaches can be faster if the dataset is not "too big". In addition, there can be trade-offs between using one vs. another hardware component more efficiently. Hence, using these strategies can be tricky, and the best approach might well depend on the specific situation. The aim is thus to make you comfortable with answering the question, "How can I use my local computing environment more efficiently to further speed up this specific analytics task?"

4. *Scaling up and scaling out*: once you have properly considered all of the above, but the task still cannot be done in a reasonable amount of time, you will need to either *scale up* or *scale out* the available computing resources. *Scaling up* refers to enlarging your machine (e.g., adding more random access memory) or switching to a more powerful machine altogether. Technically, this can mean literally building an additional hardware device into your PC; today it usually means renting a virtual server in the cloud. Instead of using a "bigger machine", *scaling out* means using several machines in concert (cluster computer, distributed systems). While this also has often been done locally (connecting several PCs to a cluster of PCs to combine all their computing power), today this too is usually done in the cloud (due to the much easier set up and maintenance). Practically, a key difference between scaling out and scaling up is that by-and-large scaling up does not require you to get familiar with specialized

software. You can simply run the exact same script you tested locally on a larger machine in the cloud. Although most of the tools and services available to scale out your analyses are by now also quite easy to use, you will have to get familiar with some additional software components to really make use of the latter. In addition, in some situations, scaling up might be perfectly sufficient, while in others only scaling out makes sense (particularly if you need massive amounts of memory). In any event, you should be comfortable dealing with the questions, "Does it make sense to scale up or scale out?" and "If yes, how can it be done?" in a given situation.[2]

Whether one or the other approach is "better" is sometimes a topic hotly debated between academics and/or practitioners with different academic backgrounds. The point of the following chapters is not to argue for one or the other approach, but to make you familiar with these different perspectives in order to make you more comfortable and able to take on large amounts of data for your analytics project. When might one or the other approach/perspective be more useful? This is highly context-dependent. However, as a general rule of thumb, consider the order in which the different approaches have been presented above.

- First, ask yourself whether there isn't an absolutely trivial solution to your big N problem, such as taking a random sample. I know, this sound banal, and you would be surprised at how many books and lectures focusing on the data engineering side of big N do not even mention this. But, we should not forget that the entire apparatus of statistical inference is essentially based on this idea.[3] There is, however, a well-justified excuse for not simply taking a random sample of a large dataset. Both in academic research and in business data science and business analytics, the decision to be facilitated with data might in any event only have measurable consequences in rather a few cases. That is, the effect size of deciding either for A or B is anyway expected to be small, and hence we need sufficient statistical power (large N) to make a meaningful decision.

- Second, once you know which statistical procedure should be run on which final sample/dataset, be aware of how to write your analytics scripts in the most efficient way. As you will see in Chapter 4, there are a handful of R idiosyncrasies

---

[2]Importantly, the perspective on scaling up and scaling out provided in this book is solely focused on Big Data Analytics in the context of economic/business research. There is a large array of practical problems and corresponding solutions/tools to deal with "Big Data Analytics" in the context of application development (e.g. tools related to data streams), which this book does not cover.

[3]Originally, one could argue, the motivation for the development of statistical inference was rather related to the practical problem of gathering data on an entire population than handling a large dataset with observations of the entire population. However, in practice, inferring population properties from a random sample also works for the latter.

that are worth keeping in mind in this regard. This will make interactive sessions in the early, exploratory phase of a Big Data project much more comfortable.

- Third, once you have a clearer idea of the bottlenecks in the data preparation and analytics scripts, aim to optimize the usage of the available local computing resources.

- In almost any organizational structure, be it a university department, a small firm, or a multinational conglomerate, switching from your laptop or desktop computer to a larger computing infrastructure, either locally or in the cloud, means additional administrative and budgetary hurdles (which means money and time spent on something other than interpreting data analysis results). That is, even before setting up the infrastructure and transferring your script and data, you will have to make an effort to scale up or scale out. Therefore, as a general rule of thumb, this option will be considered as a measure of last resort in this book.

Following this recommended order of consideration, before we focus extensively on the topics of *using local computing resources more efficiently* and *scaling up/out* (in parts II and III of this book, respectively), we need to establish some of the basics regarding what is meant by statistical/econometric solutions for big P and big N problems (in the next chapter), as well as introducing a couple of helpful programming tools and skills for working on computationally intense tasks (in Chapter 4).

3

The Two Domains of Big Data Analytics

As discussed in the previous chapter, data analytics in the context of Big Data can be broadly categorized into two domains of statistical challenges: techniques/estimators to address *big P* problems and techniques/estimators to address *big N* problems. While this book predominantly focuses on how to handle Big Data for applied economics and business analytics settings in the context of *big N* problems, it is useful to set the stage for the following chapters with two practical examples concerning both *big P* and *big N* methods.

3.1   A practical *big P* problem

Due to the abundance of digital data on all kinds of human activities, both empirical economists and business analysts are increasingly confronted with high-dimensional data (many signals, many variables). While having a lot of variables to work with sounds kind of like a good thing, it introduces new problems in coming up with useful predictive models. In the extreme case of having more variables in the model than observations, traditional methods cannot be used at all. In the less extreme case of just having dozens or hundreds of variables in a model (and plenty of observations), we risk "falsely" discovering seemingly influential variables and consequently coming up with a model with potentially very misleading out-of-sample predictions. So how can we find a reasonable model?[1] Let us look at a real-life example. Suppose you work for Google's e-commerce platform www.googlemerchandisestore.com, and you are in charge of predicting purchases (i.e., the probability that a user actually buys something from your store in a

---

[1]Note that finding a model with good in-sample prediction performance is trivial when you have a lot of variables: simply adding more variables will improve the performance. However, that will inevitably result in a nonsensical model as even highly significant variables might not have any actual predictive power when looking at out-of-sample predictions. Hence, in this kind of exercise we should *exclusively focus on out-of-sample predictions* when assessing the performance of candidate models.

given session) based on user and browser-session characteristics.[2] The dependent variable purchase is an indicator equal to 1 if the corresponding shop visit leads to a purchase and equal to 0 otherwise. All other variables contain information about the user and the session (Where is the user located? Which browser is (s)he using? etc.).

3.1.1 Simple logistic regression (naive approach)

As the dependent variable is binary, we will first estimate a simple logit model, in which we use the origins of the store visitors (how did a visitor end up in the shop?) as explanatory variables. Note that many of these variables are categorical, and the model matrix thus contains a lot of "dummies" (indicator variables). The plan in this (intentionally naive) first approach is to simply add a lot of explanatory variables to the model, run logit, and then select the variables with statistically significant coefficient estimates as the final predictive model. The following code snippet covers the import of the data, the creation of the model matrix (with all the dummy-variables), and the logit estimation.

```
# import/inspect data
ga <- read.csv("data/ga.csv")
head(ga[, c("source", "browser", "city", "purchase")])
```

```
##        source browser          city purchase
## 1      google  Chrome      San Jose        1
## 2    (direct)    Edge     Charlotte        1
## 3    (direct)  Safari San Francisco        1
## 4    (direct)  Safari   Los Angeles        1
## 5    (direct)  Chrome       Chicago        1
## 6    (direct)  Chrome     Sunnyvale        1
```

```
# create model matrix (dummy vars)
mm <- cbind(ga$purchase,
            model.matrix(purchase~source, data=ga,)[,-1])
mm_df <- as.data.frame(mm)
# clean variable names
names(mm_df) <- c("purchase",
                  gsub("source", "", names(mm_df)[-1]))
```

---

[2]We will in fact be working with a real-life Google Analytics dataset from www.googlemerchandisestore.com (https://shop.googlemerchandisestore.com); see here for details about the dataset: https://www.blog.google/products/marketingplatform/analytics/introducing-google-analytics-sample/.

```
# run logit
model1 <- glm(purchase ~ .,
              data=mm_df, family=binomial)
```

Now we can perform the t-tests and filter out the "relevant" variables.

```
model1_sum <- summary(model1)
# select "significant" variables for final model
pvalues <- model1_sum$coefficients[,"Pr(>|z|)"]
vars <- names(pvalues[which(pvalues<0.05)][-1])
vars
```

```
##  [1] "bing"
##  [2] "dfa"
##  [3] "docs.google.com"
##  [4] "facebook.com"
##  [5] "google"
##  [6] "google.com"
##  [7] "m.facebook.com"
##  [8] "Partners"
##  [9] "quora.com"
## [10] "siliconvalley.about.com"
## [11] "sites.google.com"
## [12] "t.co"
## [13] "youtube.com"
```

Finally, we re-estimate our "final" model.

```
# specify and estimate the final model
finalmodel <- glm(purchase ~.,
              data = mm_df[, c("purchase", vars)],
              family = binomial)
```

The first problem with this approach is that we should not trust the coefficient t-tests based on which we have selected the covariates too much. The first model contains 62 explanatory variables (plus the intercept). With that many hypothesis tests, we are quite likely to reject the NULL of no predictive effect although there is actually no predictive effect. In addition, this approach turns out to be unstable. There might be correlation between some of the variables in the original set, and adding/removing even one variable might substantially affect the predictive power of the model (and the apparent relevance of other variables). We can see this already from the summary of our final model estimate (generated in the next code chunk). One of the apparently relevant predictors (dfa) is not at all significant anymore in this specification. Thus, we might be tempted to further change the model, which in turn would again change the apparent relevance of other covariates, and so on.

```
summary(finalmodel)$coef[,c("Estimate", "Pr(>|z|)")]
```

```
##                           Estimate    Pr(>|z|)
## (Intercept)                -1.3831   0.000e+00
## bing                       -1.4647   4.416e-03
## dfa                        -0.1865   1.271e-01
## docs.google.com            -2.0181   4.714e-02
## facebook.com               -1.1663   3.873e-04
## google                     -1.0149  6.321e-168
## google.com                 -2.9607   3.193e-05
## m.facebook.com             -3.6920   2.331e-04
## Partners                   -4.3747   3.942e-14
## quora.com                  -3.1277   1.869e-03
## siliconvalley.about.com    -2.2456   1.242e-04
## sites.google.com           -0.5968   1.356e-03
## t.co                       -2.0509   4.316e-03
## youtube.com                -6.9935   4.197e-23
```

An alternative approach would be to estimate models based on all possible combinations of covariates and then use that sequence of models to select the final model based on some out-of-sample prediction performance measure. Clearly such an approach would take a long time to compute.

3.1.2   Regularization: the lasso estimator

Instead, the *lasso estimator* provides a convenient and efficient way to get a sequence of candidate models. The key idea behind lasso is to penalize model complexity (the cause of instability) during the estimation procedure.[3] In a second step, we can then select a final model from the sequence of candidate models based on, for example, "out-of-sample" prediction in a k-fold cross validation. The gamlr package (Taddy, 2017) provides both parts of this procedure (lasso for the sequence of candidate models, and selection of the "best" model based on k-fold cross-validation).

---

[3]In simple terms, this is done by adding $\lambda \sum_k |\beta_k|$ as a "cost" to the optimization problem.

```
# load packages
library(gamlr)
# create the model matrix
mm <- model.matrix(purchase~source, data = ga)
```

In cases with both many observations and many candidate explanatory variables, the model matrix might get very large. Even simply generating the model matrix might be a computational burden, as we might run out of memory to hold the model matrix object. If this large model matrix is sparse (i.e, has a lot of 0 entries), there is a much more memory-efficient way to store it in an R object. R provides ways to represent such sparse matrices in a compressed way in specialized R objects (such as CsparseMatrix provided in the Matrix package Bates et al. (2022)). Instead of containing all $n \times m$ cells of the matrix, these objects only explicitly store the cells with non-zero values and the corresponding indices. Below, we make use of the high-level sparse.model.matrix function to generate the model matrix and store it in a sparse matrix object. To illustrate the point of a more memory-efficient representation, we show that the traditional matrix object is about 7.5 times larger than the sparse version.

```
# create the sparse model matrix
mm_sparse <- sparse.model.matrix(purchase~source, data = ga)
# compare the object's sizes
as.numeric(object.size(mm)/object.size(mm_sparse))
```

```
## [1] 7.525
```

Finally, we run the lasso estimation with k-fold cross-validation.

```
# run k-fold cross-validation lasso
cvpurchase <- cv.gamlr(mm_sparse, ga$purchase, family="binomial")
```

We can then illustrate the performance of the selected final model – for example, with an ROC curve. Note that both the coef method and the predict method for gamlr objects automatically select the 'best' model.

```
# load packages
library(PRROC)
# use "best" model for prediction
# (model selection based on average OSS deviance
pred <- predict(cvpurchase$gamlr, mm_sparse, type="response")
```

```
# compute tpr, fpr; plot ROC
comparison <- roc.curve(scores.class0 = pred,
                        weights.class0=ga$purchase,
                        curve=TRUE)
plot(comparison)
```

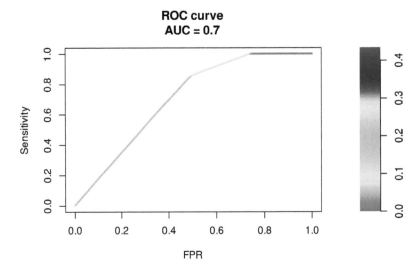

Hence, econometrics techniques such as lasso help deal with *big P* problems by providing reasonable ways to select a good predictive model (in other words, decide which of the many variables should be included).

3.2   A practical *big N* problem

Big N problems are situations in which we know what type of model we want to use but the *number of observations* is too big to run the estimation (the computer crashes or slows down significantly). The simplest statistical solution to such a problem is usually to just estimate the model based on a smaller sample. However, we might not want to do that for other reasons (i.e., if we require a big N for statistical power reasons). As an illustration of how an alternative statistical procedure can speed up the analysis of big N datasets, we look at a procedure to estimate linear models for situations where the classical OLS estimator is computationally too demanding when analyzing large datasets, the *Uluru* algorithm (Dhillon et al., 2013).

3.2.1   OLS as a point of reference

Recall the OLS estimator in matrix notation, given the linear model $\mathbf{y} = \mathbf{X}\beta + \epsilon$:

$$\hat{\beta}_{OLS} = (\mathbf{X}^\top\mathbf{X})^{-1}\mathbf{X}^\top\mathbf{y}.$$

In order to compute $\hat{\beta}_{OLS}$, we have to compute $(\mathbf{X}^\top\mathbf{X})^{-1}$, which implies a computationally expensive matrix inversion.[4] If our dataset is large, $\mathbf{X}$ is large, and the inversion can take up a lot of computation time. Moreover, the inversion and matrix multiplication to get $\hat{\beta}_{OLS}$ needs a lot of memory. In practice, it might well be that the estimation of a linear model via OLS with the standard approach in R (lm()) brings a computer to its knees, as there is not enough memory available. To further illustrate the point, we implement the OLS estimator in R.

```
beta_ols <-
    function(X, y) {
        # compute cross products and inverse
        XXi <- solve(crossprod(X,X))
        Xy <- crossprod(X, y)
        return( XXi  %*% Xy )
    }
```

Now, we will test our OLS estimator function with a few (pseudo-)random numbers in a Monte Carlo study. First, we set the sample size parameters n (the number of observations in our pseudo-sample) and p (the number of variables describing each of these observations) and initialize the dataset x.

```
# set parameter values
n <- 10000000
p <- 4
# generate sample based on Monte Carlo
# generate a design matrix (~ our 'dataset')
# with 4 variables and 10,000 observations
X <- matrix(rnorm(n*p, mean = 10), ncol = p)
# add column for intercept
X <- cbind(rep(1, n), X)
```

Now we define what the real linear model that we have in mind looks like and compute the output y of this model, given the input x.[5]

---

[4]The computational complexity of this is larger than $O(n^2)$. That is, for an input of size n, the time needed to compute (or the number of operations needed) is larger than n^2.

[5]In reality we would not know this, of course. Acting as if we knew the real model is exactly the point of Monte Carlo studies. They allow us to analyze the properties of estimators by simulation.

```
# MC model
y <- 2 + 1.5*X[,2] + 4*X[,3] - 3.5*X[,4] + 0.5*X[,5] + rnorm(n)
```

Finally, we test our `beta_ols` function.

```
# apply the OLS estimator
beta_ols(X, y)
```

```
##            [,1]
## [1,]    1.9974
## [2,]    1.5001
## [3,]    3.9996
## [4,]   -3.4994
## [5,]    0.4999
```

3.2.2   The *Uluru* algorithm as an alternative to OLS

Following Dhillon et al. (2013), we implement a procedure to compute $\hat{\beta}_{Uluru}$:

$$\hat{\beta}_{Uluru} = \hat{\beta}_{FS} + \hat{\beta}_{correct},$$

where

$$\hat{\beta}_{FS} = (\mathbf{X}_{subs}^{\top}\mathbf{X}_{subs})^{-1}\mathbf{X}_{subs}^{\top}\mathbf{Y}_{subs},$$

and

$$\hat{\beta}_{correct} = \frac{n_{subs}}{n_{rem}} \cdot (\mathbf{X}_{subs}^{\top}\mathbf{X}_{subs})^{-1}\mathbf{X}_{rem}^{\top}\mathbf{R}_{rem},$$

and

$$\mathbf{R}_{rem} = \mathbf{Y}_{rem} - \mathbf{X}_{rem} \cdot \hat{\beta}_{FS}.$$

The key idea behind this is that the computational bottleneck of the OLS estimator, the cross product and matrix inversion, $(\mathbf{X}^{\top}\mathbf{X})^{-1}$, is only computed on a subsample (X_{subs}, etc.), not the entire dataset. However, the remainder of the dataset is also taken into consideration (in order to correct a bias arising from the subsampling). Again, we implement the estimator in R to further illustrate this point.

```
beta_uluru <-
    function(X_subs, y_subs, X_rem, y_rem) {
        # compute beta_fs
        #(this is simply OLS applied to the subsample)
        XXi_subs <- solve(crossprod(X_subs, X_subs))
        Xy_subs <- crossprod(X_subs, y_subs)
        b_fs <- XXi_subs  %*% Xy_subs
```

```
        # compute \mathbf{R}_{rem}
        R_rem <- y_rem - X_rem %*% b_fs
        # compute \hat{\beta}_{correct}
        b_correct <-
            (nrow(X_subs)/(nrow(X_rem))) *
            XXi_subs %*% crossprod(X_rem, R_rem)
        # beta uluru
        return(b_fs + b_correct)
    }
```

We then test it with the same input as above:

```
# set size of sub-sample
n_subs <- 1000
# select sub-sample and remainder
n_obs <- nrow(X)
X_subs <- X[1L:n_subs,]
y_subs <- y[1L:n_subs]
X_rem <- X[(n_subs+1L):n_obs,]
y_rem <- y[(n_subs+1L):n_obs]
# apply the uluru estimator
beta_uluru(X_subs, y_subs, X_rem, y_rem)
```

```
##           [,1]
## [1,]   2.0048
## [2,]   1.4997
## [3,]   3.9995
## [4,]  -3.4993
## [5,]   0.4996
```

This looks quite good already. Let's have a closer look with a little Monte Carlo study. The aim of the simulation study is to visualize the difference between the classical OLS approach and the *Uluru* algorithm with regard to bias and time complexity if we increase the sub-sample size in *Uluru*. For simplicity, we only look at the first estimated coefficient β_1.

```
# define sub-samples
n_subs_sizes <- seq(from = 1000, to = 500000, by=10000)
n_runs <- length(n_subs_sizes)
# compute uluru result, stop time
mc_results <- rep(NA, n_runs)
```

```
mc_times <- rep(NA, n_runs)
for (i in 1:n_runs) {
    # set size of sub-sample
    n_subs <- n_subs_sizes[i]
    # select sub-sample and remainder
    n_obs <- nrow(X)
    X_subs <- X[1L:n_subs,]
    y_subs <- y[1L:n_subs]
    X_rem <- X[(n_subs+1L):n_obs,]
    y_rem <- y[(n_subs+1L):n_obs]
    mc_results[i] <- beta_uluru(X_subs,
                                y_subs,
                                X_rem,
                                y_rem)[2] # (1 is the intercept)
    mc_times[i] <- system.time(beta_uluru(X_subs,
                                          y_subs,
                                          X_rem,
                                          y_rem))[3]
}
# compute OLS results and OLS time
ols_time <- system.time(beta_ols(X, y))
ols_res <- beta_ols(X, y)[2]
```

Let's visualize the comparison with OLS.

```
# load packages
library(ggplot2)
# prepare data to plot
plotdata <- data.frame(beta1 = mc_results,
                       time_elapsed = mc_times,
                       subs_size = n_subs_sizes)
```

First, let's look at the time used to estimate the linear model.

```
ggplot(plotdata, aes(x = subs_size, y = time_elapsed)) +
    geom_point(color="darkgreen") +
    geom_hline(yintercept = ols_time[3],
               color = "red",
               linewidth = 1) +
    theme_minimal() +
```

```
ylab("Time elapsed") +
xlab("Subsample size")
```

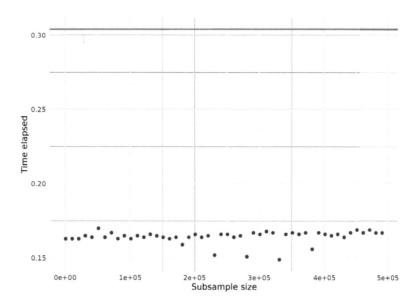

The horizontal red line indicates the computation time for estimation via OLS; the green points indicate the computation time for the estimation via the *Ulruru* algorithm. Note that even for large sub-samples, the computation time is substantially lower than for OLS. Finally, let's have a look at how close the results are to OLS.

```
ggplot(plotdata, aes(x = subs_size, y = beta1)) +
    geom_hline(yintercept = ols_res,
               color = "red",
               linewidth = 1) +
    geom_hline(yintercept = 1.5,
               color = "green",
               linewidth = 1) +
    geom_point(color="darkgreen") +
    theme_minimal() +
    ylab("Estimated coefficient") +
    xlab("Subsample size")
```

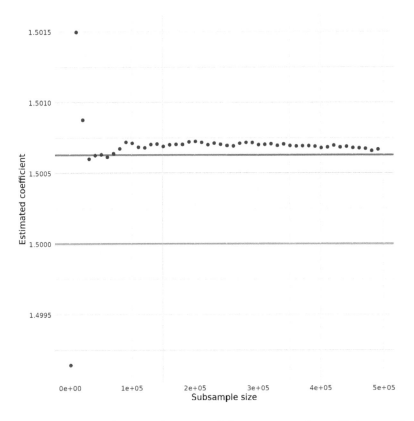

The horizontal red line indicates the size of the estimated coefficient, when using OLS. The horizontal green line indicates the size of the actual coefficient. The green points indicate the size of the same coefficient estimated by the *Uluru* algorithm for different sub-sample sizes. Note that even relatively small sub-samples already deliver estimates very close to the OLS estimates. Taken together, the example illustrates that alternative statistical methods, optimized for large amounts of data, can deliver results very close to traditional approaches. Yet, they can deliver these results much more efficiently.

Part II

Platform: Software and Computing Resources

Introduction

"Are tomorrow's bigger computers going to solve the problem? For some people, yes – their data will stay the same size and computers will get big enough to hold it comfortably. For other people it will only get worse – more powerful computers mean extraordinarily larger datasets. If you are likely to be in this latter group, you might want to get used to working with databases now." (Burns, 2011, p.16)

At the center of Figure 3.1 you see an illustration of the key components of a standard local computing environment to process digital data. In this book, these components typically serve the purpose of computing a statistic, given a large dataset as input. In this part of the book, you will get familiar with how each of these components plays a role in the context of Big Data Analytics and how you can recognize and resolve potential problems caused by large datasets/large working loads for either of these components. The three most relevant of these components are:

- *Mass storage*. This is the type of computer memory we use to store data over the long term. The mass storage device in our local computing environment (e.g., our PC/laptop) is generally referred to as the *hard drive* or *hard disk*.

- *RAM*. In order to work with data (e.g., in R), it first has to be loaded into the *memory* of our computer – more specifically, into the random access memory (*RAM*). Typically, data is only loaded in the RAM for as long as we are working with it.

- *CPU*. The component actually *processing* data is the Central Processing Unit (CPU). When using R to process data, R commands are translated into complex combinations of a small sets of basic operations, which the *CPU* then executes.

For the rest of this book, consider the main difference between common 'data analytics' and 'Big Data Analytics' to be the following: in a Big Data Analytics context, the standard usage of one or several of the standard hardware components in your local computer fails to work or works very inefficiently because the amount of data overwhelms its normal capacity.

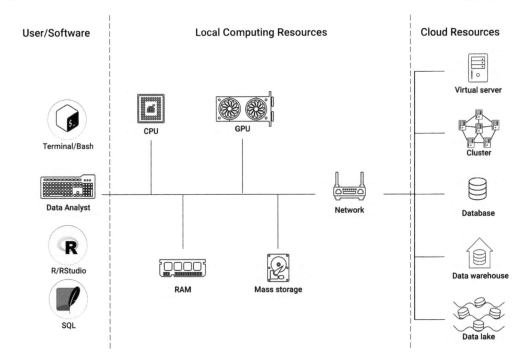

FIGURE 3.1: Software, local computing resources, and cloud resources.

However, before we discuss these *hardware* issues in Chapter 5, Chapter 4 focuses on the software we will work with to make use of these hardware components in the most efficient way. This perspective is symbolized in the left part of Figure 3.1. There will be three main software components with which we work in what follows. The first is the terminal (i.e., bash): instead of using graphical user interfaces to interact with our local computing environment, we will rather use the terminal[6] to install software, download files, inspect files, and inspect hardware performance. For those of you not yet used to working with the terminal, do not worry! There are no prerequisites in knowledge about working with the terminal, and most of the use cases in this book are very simple and well explained. At this point, just note that there will be two types of code chunks (code examples) shown in what follows: either they show code that should be run in the terminal (in RStudio, such code should thus be entered in the tab/window called *Terminal*), or *R* code (in RStudio, this code should be entered in the Tab/window called *Console*).

---

[6]To be consistent with the cloud computing services (particularly EC2) introduced later on, the sections involving terminal commands assume you work with Ubuntu Linux. However, the corrsponding code chunks are essentially identical for users working on a Mac/OSX-machine. In case you are working on a Windows machine, starting with Windows 10, you will have essentially the same tool available called the *Windows Terminal*. In older Windows versions the Linux terminal equivalent is called PowerShell or command prompt, which use a slightly different syntax, but provide similar functionality. See https://www.geeksforgeeks.org/linux-vs-windows-commands/ for an overview of how to get started with the Windows command prompt, including a detailed listing of commands next to their corresponding Linux command equivalents.

R (R Core Team, 2021) will be the main language used throughout this book. It will be the primary language to gather, import, clean, visualize, and analyze data. However, as you will already see in some of the examples in this part of the book, we will often not necessarily use base R, but rather use R as a high-level interface to more specialized software and services. On the one hand this means we will install and use several specialized R packages designed to manipulate large datasets that are in fact written in other (faster) languages, such as C. On the other hand, it will mean that we use R commands to communicate with other lower-level software systems that are particularly designed to handle large amounts of data (such as data warehouses, or software to run analytics scripts on cluster computers). The point is that even if the final computation is not actually done in R, all you need to know to get the particular job done are the corresponding R commands to trigger this final computation.

There is one more indispensable software tool on which we will build throughout the book, Structured Query Language (SQL), or to be precise, different variants of SQL in different contexts. The main reason for this is twofold: First, even if you primarily interact with some of the lower-level Big Data software tools from within R, it is often more comfortable (or, in some cases, even necessary) to send some of the instructions in the form of SQL commands (wrapped in an R function); second, while many of you might have only heard of SQL in the context of traditional relational database systems, SQL variants are nowadays actually used to interact with a variety of the most important Big Data systems, ranging from Apache Spark (a unified analytics platform for large-scale data processing) to Apache Druid (a column-based distributed data store) and AWS Athena (a cloud-based, serverless query service for simple storage/data lakes). Hence, if you work in data analytics/-data science and want to seriously engage with Big Data, knowing your way around SQL is a very important asset.

Finally, Chapters 6 and 7 consider the situation where all of the tweaks to use the local computing resources most efficiently are still not enough to get the job done (one or several components are still overwhelmed and/or it simply takes too much time to run the analysis). From the hardware perspective, there are two basic strategies to cope with such a situation:

- *Scale out ('horizontal scaling')*: Distribute the workload over several computers (or separate components of a system).
- *Scale up ('vertical scaling')*: Extend the physical capacity of the affected component by building a system with a large amount of RAM shared between applications. This sounds like a trivial solution ('if RAM is too small, buy more RAM…'), but in practice it can be very expensive.

Nowadays, either of these approaches is typically taken with the help of cloud resources (illustrated in the right part of Figure 3.1). How this is basically done is introduced in Chapter 7. Regarding *vertical scaling*, you will see that the transition from a local computing environment to the cloud (involving some or all of the core computing components) is rather straightforward to learn. However, *horizontal scaling* for really massive datasets involves some new hardware and software concepts related to what are generally called *distributed systems*. To this end, we first introduce the most relevant concepts related to distributed systems in Chapter 6.

4

Software: Programming with (Big) Data

The programming language and computing environment R (R Core Team, 2021) is particularly made for writing code in a data analytics context. However, the language was developed at a time when data analytics was primarily focused on moderately sized datasets that can easily be loaded/imported and worked with on a common PC. Depending on the field or industry you work in, this is not the case anymore. In this chapter, we will explore some of R's (potential) weaknesses as well as learn how to avoid them and how to exploit some of R's strengths when it comes to working with large datasets. The first part of this chapter is primarily focused on understanding code profiling and improving code with the aim of making computationally intensive data analytics scripts in R run faster. This chapter presupposes basic knowledge of R data structures and data types as well as experience with basic programming concepts such as loops.[1] While very useful in writing analytics scripts, we will not look into topics like coding workflows, version control, and code sharing (e.g., by means of Git and GitHub[2]). The assumption is that you bring some experience in writing analytics scripts already.

While R is a very useful tool for many aspects of Big Data Analytics that we will cover in the following chapters, R alone is not enough for a basic Big Data Analytics toolbox. The second part of this chapter introduces the reader to the *Structured Query Language (SQL)*, a programming language designed for managing data in relational databases. Although the type of databases where SQL is traditionally encountered would not necessarily be considered part of Big Data Analytics today, some versions of SQL are now used with systems particularly designed for Big Data Analytics (such as Amazon Athena and Google BigQuery). Hence, with a good knowledge of R in combination with basic SQL skills, you will be able to productively engage with a large array of practical Big Data Analytics problems.

---

[1]Appendix B reviews the most relevant concepts regarding data types and data structures in R.
[2]However, in case you want to further explore how to use GitHub as part of your coding workflow, Appendix A provides a very short introduction to the topic.

4.1   Domains of programming with (big) data

Programming tasks in the context of data analytics typically fall into one of the following broad categories:

- Procedures to import/export data.
- Procedures to clean and filter data.
- Implementing functions for statistical analysis.

When writing a program to process large amounts of data in any of these areas, it is helpful to take into consideration the following design choices:

1. Which basic (already implemented) R functions are more or less suitable as building blocks for the program?[3]
2. How can we exploit/avoid some of R's lower-level characteristics in order to write more efficient code?
3. Is there a need to interface with a lower-level programming language in order to speed up the code? (advanced topic)

Finally, there is an additional important point to be made regarding the writing of code for *statistical analysis*: Independent of *how* we write a statistical procedure in R (or in any other language, for that matter), keep in mind that there might be an *alternative statistical procedure/algorithm* that is faster but delivers approximately the same result (as long as we use a sufficiently large sample).

4.2   Measuring R performance

When writing a data analysis script in R to process large amounts of data, it generally makes sense to first test each crucial part of the script with a small sub-sample. In order to quickly recognize potential bottlenecks, there are a couple of R packages that help you keep track of exactly how long each component of your script needs to process as well as how much memory it uses. The table below lists some of the packages and functions that you should keep in mind when *"profiling"* and testing your code.

---

[3]Throughout the rest of this book, I will point to specialized R packages and functions that are particularly designed to work with large amounts of data. Where necessary, we will also look more closely at the underlying concepts that explain why these specialized packages work better with large amounts of data than the standard approaches.

package	function	purpose
utils	object.size()	Provides an estimate of the memory that is being used to store an R object.
pryr	object_size()	Works similarly to object.size(), but counts more accurately and includes the size of environments.
pryr	mem_used()	Returns the total amount of memory (in megabytes) currently used by R.
pryr	mem_change()	Shows the change in memory (in megabytes) before and after running code.
base	system.time()	Returns CPU (and other) times that an R expression used.
microbenchmark	microbenchmark()	Highly accurate timing of R expression evaluation.
bench	mark()	Benchmark a series of functions.
profvis	profvis()	Profiles an R expression and visualizes the profiling data (usage of memory, time elapsed, etc.).

Most of these functions are used in an interactive way in the R console. They serve either of two purposes that are central to profiling and improving your code's performance. First, in order to assess the performance of your R code you probably want to know how long it takes to run your entire script or a specific part of your script. The system.time() (R Core Team, 2021) function provides an easy way to check this. This function is loaded by default with R; there is no need to install an additional package. Simply wrap it around the line(s) of code that you want to assess.

```r
# how much time does it take to run this loop?
system.time(for (i in 1:100) {i + 5})
```

```
##    user  system elapsed
##   0.002   0.000   0.002
```

Note that each time you run this line of code, the returned amount of time varies slightly. This has to do with the fact that the actual time needed to run a line of code can depend on various other processes happening at the same time on your computer.

The microbenchmark (Mersmann, 2021) and bench (Hester and Vaughan, 2021) packages provide additional functions to measure execution time in more sophisticated ways. In particular, they account for the fact that the processing time for the same code might vary and automatically run the code several times in order to return statistics about the processing time. In addition, `microbenchmark()` provides highly detailed and highly accurate timing of R expression evaluation. The function is particularly useful to accurately find even minor room for improvement when testing a data analysis script on a smaller sub-sample (which might scale when working on a large dataset). For example, suppose you need to run a for-loop over millions of iterations, and there are different ways to implement the body of the loop (which does not take too much time to process in one iteration). Note that the function actually evaluates the R expression in question many times and returns a statistical summary of the timings.

```r
# load package
library(microbenchmark)
# how much time does it take to run this loop (exactly)?
microbenchmark(for (i in 1:100) {i + 5})
```

```
## Unit: milliseconds
##                            expr    min     lq   mean
##   for (i in 1:100) {    i + 5 } 1.116  1.231  1.308
##   median    uq    max neval
##    1.265 1.315 4.838    100
```

Second, a key aspect to improving the performance of data analysis scripts in R is to detect inefficient memory allocation as well as avoiding an R-object that is either growing too much or too large to handle in memory. To this end, you might want to monitor how much memory R occupies at different points in your script as well as how much memory is taken up by individual R objects. For example, `object.size()` returns the size of an R object, that is, the amount of memory it takes up in the R environment in bytes (`pryr::object_size()` counts slightly more accurately).

```r
hello <- "Hello, World!"
object.size(hello)
```

```
## 120 bytes
```

This is useful to implementing your script with a generally less memory-intensive approach. For example, for a specific task it might not matter whether a particular variable is stored as a `character` vector or a `factor`. But storing it as `character` turns out to be more memory intensive (why?).

```
# initialize a large string vector containing letters
large_string <- rep(LETTERS[1:20], 1000^2)
head(large_string)
```

```
## [1] "A" "B" "C" "D" "E" "F"
```

```
# store the same information as a factor in a new variable
large_factor <- as.factor(large_string)
```

```
# is one bigger than the other?
object.size(large_string) - object.size(large_factor)
```

```
## 79999456 bytes
```

`pryr::mem_change()` (Wickham, 2021) is useful to track how different parts of your script affect the overall memory occupied by R.

```
# load package
library(pryr)
```

```
# initialize a vector with 1000 (pseudo)-random numbers
mem_change(
        thousand_numbers <- runif(1000)
        )
```

```
## 7.98 kB
```

```
# initialize a vector with 1M (pseudo)-random numbers
mem_change(
        a_million_numbers <- runif(1000^2)
        )
```

```
## 8 MB
```

`bench::mark()` allows you to easily compare the performance of several different implementations of a code chunk both regarding timing and memory usage. The following code example illustrates this in a comparison of two approaches to computing the product of each element in a vector x with a factor z.

```r
# load packages
library(bench)


# initialize variables
x <- 1:10000
z <- 1.5


# approach I: loop
multiplication <-
        function(x,z) {
                result <- c()
                for (i in 1:length(x)) {result <- c(result, x[i]*z)}
                return(result)
        }
result <- multiplication(x,z)
head(result)


## [1] 1.5 3.0 4.5 6.0 7.5 9.0


# approach II: "R-style"
result2 <- x * z
head(result2)


## [1] 1.5 3.0 4.5 6.0 7.5 9.0


# comparison
benchmarking <-
        mark(
        result <- multiplication(x,z),
        result2 <- x * z,
        min_iterations = 50
)
benchmarking[, 4:9]


## # A tibble: 2 x 3
##    `itr/sec` mem_alloc `gc/sec`
##        <dbl> <bch:byt>    <dbl>
## 1      12.8      382MB     17.7
## 2   70057.     78.2KB      7.01
```

In addition, the bench package (Hester and Vaughan, 2021) provides a simple way to visualize these outputs:

```
plot(benchmarking, type = "boxplot")
```

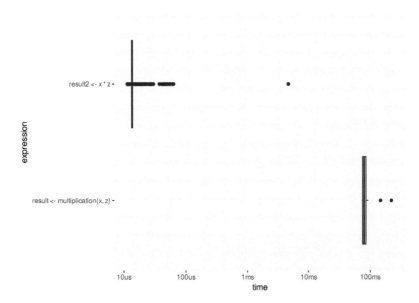

Finally, to analyze the performance of your entire script/program, the profvis package (Chang et al., 2020) provides visual summaries to quickly detect the most prominent bottlenecks. You can either call this via the profvis() function with the code section to be profiled as argument, or via the RStudio user interface by clicking on the Code Tools menu in the editor window and selecting "Profile selected lines".

```
# load package
library(profvis)

# analyze performance of several lines of code
profvis({
        x <- 1:10000
        z <- 1.5

# approach I: loop
multiplication <-
        function(x,z) {
                result <- c()
                for (i in 1:length(x)) {result <- c(result, x[i]*z)}
                return(result)
```

```
        }
result <- multiplication(x,z)

# approach II: "R-style"
result2 <- x * z
head(result2)
})
```

4.3 Writing efficient R code

This subsection touches upon several prominent aspects of writing efficient/fast R code.[4]

4.3.1 Memory allocation and growing objects

R tends to "grow" already-initialized objects in memory when they are modified. At the initiation of the object, a small amount of memory is occupied at some location in memory. In simple terms, once the object grows, it might not have enough space where it is currently located. Hence, it needs to be "moved" to another location in memory with more space available. This moving, or "re-allocation" of memory, needs time and slows down the overall process.

This potential is most practically illustrated with a for-loop in which each iteration's result is stored as an element of a vector (the object in question). To avoid growing this object, you need to instruct R to pre-allocate the memory necessary to contain the final result. If we don't do this, each iteration of the loop causes R to re-allocate memory because the number of elements in the vector/list is changing. In simple terms, this means that R needs to execute more steps in each iteration.

In the following example, we compare the performance of two functions, one taking this principle into account, the other not. The functions take a numeric vector as input and return the square root of each element of the numeric vector.

```
# naïve implementation
sqrt_vector <-
        function(x) {
```

---

[4]This is not intended to be a definitive guide to writing efficient R code in every aspect. Instead the subsection aims at covering most of the typical pitfalls to avoid and to provide a number of easy-to-remember tricks to keep in mind when writing R code for computationally intensive tasks.

```
        output <- c()
        for (i in 1:length(x)) {
            output <- c(output, x[i]^(1/2))
        }

        return(output)
    }


# implementation with pre-allocation of memory
sqrt_vector_faster <-
    function(x) {
        output <- rep(NA, length(x))
        for (i in 1:length(x)) {
            output[i] <-  x[i]^(1/2)
        }

        return(output)
    }
```

As a proof of concept we use `system.time()` to measure the difference in speed for various input sizes.[5]

```
# the different sizes of the vectors we will put into the two functions
input_sizes <- seq(from = 100, to = 10000, by = 100)
# create the input vectors
inputs <- sapply(input_sizes, rnorm)

# compute outputs for each of the functions
output_slower <-
    sapply(inputs,
        function(x){ system.time(sqrt_vector(x))["elapsed"]
            }
        )
output_faster <-
    sapply(inputs,
        function(x){ system.time(sqrt_vector_faster(x))["elapsed"]
            }
        )
```

---

[5]We generate the numeric input by drawing vectors of (pseudo-)random numbers via `rnorm()`.

The following plot shows the difference in the performance of the two functions.

```r
# load packages
library(ggplot2)

# initialize data frame for plot
plotdata <- data.frame(time_elapsed = c(output_slower, output_faster),
                       input_size = c(input_sizes, input_sizes),
                       Implementation= c(rep("sqrt_vector",
                                             length(output_slower)),
                                         rep("sqrt_vector_faster",
                                             length(output_faster))))

# plot
ggplot(plotdata, aes(x=input_size, y= time_elapsed)) +
    geom_point(aes(colour=Implementation)) +
    theme_minimal(base_size = 18) +
    theme(legend.position = "bottom") +
    ylab("Time elapsed (in seconds)") +
    xlab("No. of elements processed")
```

Clearly, the version with pre-allocation of memory (avoiding growing an object) is much faster overall. In addition, we see that the problem with the growing object in the naïve implementation tends to get worse with each iteration. The take-away message for the practitioner: If possible, always initialize the "container" object (list, matrix, etc.) for iteration results as an empty object of the final size/dimensions.

The attentive reader and experienced R coder will have noticed by this point that both of the functions implemented above are not really smart practice to solve the problem at hand. If you consider yourself part of this group, the next subsection will make you more comfortable.

4.3.2 Vectorization in basic R functions

We can further improve the performance of this function by exploiting a particular characteristic of R: in R, 'everything is a vector', and many of the most basic R functions (such as math operators) are *vectorized*. In simple terms, this means that an operation is implemented to directly work on vectors in such a way that it can take advantage of the similarity of each of the vector's elements. That is, R only has to figure out once how to apply a given function to a vector element in order to apply it to all elements of the vector. In a simple loop, however, R has to go through the same 'preparatory' steps again and again in each iteration.

Following up on the problem from the previous subsection, we implement an additional function called `sqrt_vector_fastest` that exploits the fact that math operators in R are vectorized functions. We then re-run the same speed test as above with this function.

```
# implementation with vectorization
sqrt_vector_fastest <-
    function(x) {
            output <-  x^(1/2)
        return(output)
    }


# speed test
output_fastest <-
    sapply(inputs,
            function(x){ system.time(sqrt_vector_fastest(x))["elapsed"]
                }
            )
```

Let's have a look at whether this improves the function's performance further.

```
# load packages
library(ggplot2)

# initialize data frame for plot
plotdata <- data.frame(time_elapsed = c(output_faster, output_fastest),
```

```
                    input_size = c(input_sizes, input_sizes),
                 Implementation= c(rep("sqrt_vector_faster",
                                    length(output_faster)),
                               rep("sqrt_vector_fastest",
                                  length(output_fastest))))

# plot
ggplot(plotdata, aes(x=time_elapsed, y=Implementation)) +
    geom_boxplot(aes(colour=Implementation),
                      show.legend = FALSE) +
    theme_minimal(base_size = 18) +
    xlab("Time elapsed (in seconds)")
```

Clearly, the vectorized implementation is even faster. The take-away message: Make use of vectorized basic R functions where possible. At this point you might wonder: Why not always use vectorization over loops, when working with R? This question (and closely related similar questions) has been fiercely debated in the R online community over the last few years. Also the debate contains and has contained several (in my view) slightly misleading arguments. A simple answer to this question is: It is in fact not that simple to use *actual* vectorization for every kind of problem in R. There are a number of functions often mentioned to achieve "vectorization" easily in R; however, they do not actually implement actual vectorization in its original technical sense (the type just demonstrated here with the R math operators). Since this point is very prominent in debates about how to improve R code, the next subsection attempts to summarize the most important aspects to keep in mind.

4.3.3 `apply`-type functions and vectorization

There are basically two ways to make use of some form of "vectorization" instead of writing loops.

One approach is to use an `apply`-type function instead of loops. Note, though, that the `apply`-type functions primarily make the writing of code more efficient. They still run a loop under the hood. Nevertheless, some `apply`-type functions might still outperform explicit loops, as they might be better implemented.[6]

Consider, for example, `lapply()`, a function that takes a vector (atomic or list) as input and applies a function `FUN` to each of its elements. It is a straightforward alternative to `for`-loops in many situations (and it automatically takes care of the "growing objects" problem discussed above). The following example shows how we can get the same result by either writing a loop or using `lapply()`. The aim of the code example is to import the Health News in Twitter Dataset[7] by Karami et al. (2017). The raw data consists of several text files that need to be imported to R consecutively.

The text-files are located in `data/twitter_texts/`. For either approach of importing all of these files, we first need a list of the paths to all of the files. We can get this with `list.files()`. Also, for either approach we will make use of the `fread` function in the `data.table` package (Dowle and Srinivasan, 2022).

```
# load packages
library(data.table)


# get a list of all file-paths
textfiles <- list.files("data/twitter_texts", full.names = TRUE)
```

Now we can read in all the text files with a `for`-loop as follows.

```
# prepare loop
all_texts <- list()
n_files <- length(textfiles)
length(all_texts) <- n_files
# read all files listed in textfiles
for (i in 1:n_files) {
```

---

[6]If you know how to implement efficient for-loops in R (as you are certainly expected to at this point), there is not much to gain from using an `apply`-type function instead of a loop, apart from making your code easier to read (and faster to write).

[7]https://archive.ics.uci.edu/ml/datasets/Health+News+in+Twitter

```
        all_texts[[i]] <- fread(textfiles[i])
}
```

The imported files are now stored as data.table-objects in the list all_texts. With the following line of code we combine all of them in one data.table.

```
# combine all in one data.table
twitter_text <- rbindlist(all_texts)
# check result
dim(twitter_text)
```

```
## [1] 42422    3
```

Alternatively, we can make use of lapply as follows in order to achieve exactly the same.

```
# use lapply instead of loop
all_texts <- lapply(textfiles, fread)
# combine all in one data.table
twitter_text <- rbindlist(all_texts)
# check result
dim(twitter_text)
```

```
## [1] 42422    3
```

Finally, we can make use of Vectorization() in order to "vectorize" our own import function (written for this example). Again, this does not make use of vectorization in its original technical sense.

```
# initialize the import function
import_file <-
    function(x) {
        parsed_x <- fread(x)
        return(parsed_x)
    }


# 'vectorize' it
import_files <- Vectorize(import_file, SIMPLIFY = FALSE)

# Apply the vectorized function
all_texts <- import_files(textfiles)
```

```
twitter_text <- rbindlist(all_texts)
# check the result
dim(twitter_text)
```

```
## [1] 42422     3
```

The take-away message: Instead of writing simple loops, use `apply`-type functions to save time writing code (and make the code easier to read) and automatically avoid memory-allocation problems.

4.3.4 Avoiding unnecessary copying

The "growing objects" problem discussed above is only one aspect that can lead to inefficient use of memory when working with R. Another potential problem of using up more memory than necessary during an execution of an R-script is how R handles objects/variables and their names.

Consider the following line of code:

```
a <- runif(10000)
```

What is usually said to describe what is happening here is something along the lines of "we initialize a variable called `a` and assign a numeric vector with 10,000 random numbers. What in fact happens is that the name `a` is assigned to the integer vector (which in turn exists at a specific memory address). Thus, values do not have names but *names have values*. This has important consequences for memory allocation and performance. For example, because `a` is in fact just a name attached to a value, the following does not involve any copying of values. It simply "binds" another name, `b`, to the same value to which `a` is already bound.

```
b <- a
```

We can prove this in two ways. First, if what I just stated was not true, the line above would actually lead to more memory being occupied by the current R session. However, this is not the case:

```
object_size(a)
```

```
## 80.05 kB
```

```
mem_change(c <- a)
```

```
## -588 kB
```

Second, by means of the lobstr-package (Wickham, 2022a), we can see that the values to which a and b are bound are stored at the same memory address. Hence, they are the same values.

```
# load packages
library(lobstr)
```

```
# check memory addresses of objects
obj_addr(a)
```

```
## [1] "0x55ca2050a870"
```

```
obj_addr(b)
```

```
## [1] "0x55ca2050a870"
```

Now you probably wonder, what happens to b if we modify a. After all, if the values to which b is bound are changed when we write code concerning a, we might end up with very surprising output. The answer is, and this is key (!), once we modify a, the values need to be *copied* in order to ensure the integrity of b. Only at this point, our program will require more memory.

```
# check the first element's value
a[1]
```

```
## [1] 0.5262
```

```
b[1]
```

```
## [1] 0.5262
```

```
# modify a, check memory change
mem_change(a[1] <- 0)
```

```
## 79 kB
```

```
# check memory addresses
obj_addr(a)
```

```
## [1] "0x55c9f91c8380"
```

```
obj_addr(b)
```

```
## [1] "0x55ca2050a870"
```

Note that the entire vector needed to be copied for this. There is, of course, a lesson from all this regarding writing efficient code. Knowing how actual copying of values occurs helps avoid unnecessary copying. The larger an object, the more time it will take to copy it in memory. Objects with a single binding get modified in place (no copying):

```
mem_change(d <- runif(10000))
```

```
## 80.3 kB
```

```
mem_change(d[1] <- 0)
```

```
## 584 B
```

4.3.5   Releasing memory

Closely related to the issue of copy-upon-modify is the issue of "releasing" memory via "garbage collection". If your program uses up a lot of (too much) memory (typical for working with large datasets), all processes on your computer might substantially slow down (we will look more closely into why this is the case in the next chapter). Hence, you might want to remove/delete an object once you do not need it anymore. This can be done with the rm() function.

```
mem_change(large_vector <- runif(10^8))
```

```
## 800 MB
```

```
mem_change(rm(large_vector))
```

```
## -800 MB
```

rm() removes objects that are currently accessible in the global R environment.

However, some objects/values might technically not be visible/accessible anymore (for example, objects that have been created in a function which has since returned the function output). To also release memory occupied by these objects, you can call gc() (the garbage collector). While R will automatically collect the garbage once it is close to running out of memory, explicitly calling gc can still improve the performance of your script when working with large datasets. This is in particular the case when R is not the only data-intensive process running on your computer. For example, when running an R script involving the repeated querying of data from a local SQL database and the subsequent memory-intensive processing of this data in R, you can avoid using up too much memory by running rm and gc explicitly.[8]

4.3.6   Beyond R

So far, we have explored idiosyncrasies of R we should be aware of when writing programs to handle and analyze large datasets. While this has shown that R has many advantages for working with data, it also revealed some aspects of R that might result in low performance compared to other programming languages. A simple generic explanation for this is that R is an interpreted language, meaning that when we execute R code, it is processed (statement by statement) by an 'interpreter' that translates the code into machine code (without the user giving any specific instructions). In contrast, when writing code in a 'compiled language', we first have to explicitly compile the code (into machine code) and then run the compiled program. Running code that is already compiled is typically much faster than running R code that has to be interpreted before it can actually be processed by the CPU.

For advanced programmers, R offers various options to directly make use of compiled programs (for example, written in C, C++, or FORTRAN). In fact, several of the core R functions installed with the basic R distribution are implemented in one of these lower-level programming languages, and the R function we call simply interacts with these functions.

We can actually investigate this by looking at the source code of an R function. If you simply type the name of a function (such as our import_file()) to the console, R prints the function's source code to the console.

```
import_file
```

```
## function(x) {
##          parsed_x <- fread(x)
##          return(parsed_x)
```

---

[8]Note that running gc() takes some time, so you should not overdo it. As a rule of thumb, run gc() after removing a really large object.

```
##       }
## <bytecode: 0x55ca1fdc2030>
```

However, if we do the same for function `sum`, we don't see any actual source code.

```
sum
```

```
## function (..., na.rm = FALSE)  .Primitive("sum")
```

Instead `.Primitive()` indicates that `sum()` is actually referring to an internal function (in this case implemented in C).

While the use of functions implemented in a lower-level language is a common technique to improve the speed of 'R' functions, it is particularly prominent in the context of functions/packages made to deal with large amounts of data (such as the `data.table` package).

4.4   SQL basics

Structured Query Language (SQL) has become a bread-and-butter tool for data analysts and data scientists due to its broad application in systems used to store large amounts of data. While traditionally only encountered in the context of structured data stored in relational database management systems, some versions of it are now also used to query data from data warehouse systems (e.g. Amazon Redshift) and even to query massive amounts (terabytes or even petabytes) of data stored in data lakes (e.g., Amazon Athena). In all of these applications, SQL's purpose (from the data analytics perspective) is to provide a convenient and efficient way to query data from mass storage for analysis. Instead of importing a CSV file into R and then filtering it in order to get to the analytic dataset, we use SQL to express how the analytic dataset should look (which variables and rows should be included).

The latter point is very important to keep in mind when already having experience with a language like R and learning SQL for the first time. In R we write code to instruct the computer what to do with the data. For example, we tell it to import a csv file called `economics.csv` as a `data.table`; then we instruct it to remove observations that are older than a certain date according to the `date` column; then we instruct it to compute the average of the `unemploy` column values for each year based on the `date` column and then return the result as a separate data frame:

```
# import data
econ <- read.csv("data/economics.csv")
```

```r
# filter
econ2 <- econ["1968-01-01"<=econ$date,]

# compute yearly averages (basic R approach)
econ2$year <- lubridate::year(econ2$date)
years <- unique(econ2$year)
averages <-
    sapply(years, FUN = function(x){
        mean(econ2[econ2$year==x,"unemploy"])
        })
output <- data.frame(year=years, average_unemploy=averages)

# inspect the first few lines of the result
head(output)
```

```
##   year average_unemploy
## 1 1968             2797
## 2 1969             2830
## 3 1970             4127
## 4 1971             5022
## 5 1972             4876
## 6 1973             4359
```

In contrast, when using SQL we write code that describes what the final result is supposed to look like. The SQL engine processing the code then takes care of the rest and returns the result in the most efficient way.[9]

```sql
SELECT
strftime('%Y', `date`) AS year,
AVG(unemploy) AS average_unemploy
FROM econ
WHERE "1968-01-01"<=`date`
GROUP BY year LIMIT 6;
```

```
##   year average_unemploy
## 1 1968             2797
## 2 1969             2830
```

---

[9]In particular, the user does not need to explicitly instruct SQL at which point in the process which part (filtering, selecting variables, aggregating, creating new variables etc.) of the query should be processed. SQL will automatically find the most efficient way to process the query.

```
## 3  1970            4127
## 4  1971            5022
## 5  1972            4876
## 6  1973            4359
```

For the moment, we will only focus on the code and ignore the underlying hardware and database concepts (those will be discussed in more detail in Chapter 5).

4.4.1   First steps in SQL(ite)

In order to get familiar with coding in SQL, we work with a free and easy-to-use version of SQL called *SQLite*. SQLite[10] is a free full-featured SQL database engine widely used across platforms. It usually comes pre-installed with Windows and Mac/OSX distributions and has (from the user's perspective) all the core features of more sophisticated SQL versions. Unlike the more sophisticated SQL systems, SQLite does not rely explicitly on a client/server model. That is, there is no need to set up your database on a server and then query it from a client interface. In fact, setting it up is straightforward. In the terminal, we can directly call SQLite as a command-line tool (on most modern computers, the command is now `sqlite3`, SQLite version 3).

In this first code example, we set up an SQLite database using the command line. In the file structure of the book repository, we first switch to the data directory.

```
cd data
```

With one simple command, we start up SQLite, create a new database called `mydb.sqlite`, and connect to the newly created database.[11]

```
sqlite3 mydb.sqlite
```

This created a new file `mydb.sqlite` in our `data` directory, which contains the newly created database. Also, we are now running `sqlite` in the terminal (indicated by the `sqlite>` prompt. This means we can now type SQL code to the terminal to run queries and other SQL commands.

At this point, the newly created database does not contain any data. There are no tables in it. We can see this by running the `.tables` command.

---

[10]https://sqlite.org/index.html

[11]If there is already a database called `mydb.sqlite` in this folder, the same command would simply start up SQLite and connect to the existing database.

```
.tables
```

As expected, nothing is returned. Now, let's create our first table and import the economics.csv dataset into it. In SQLite, it makes sense to first set up an empty table in which all column data types are defined before importing data from a CSV-file to it. If a CSV is directly imported to a new table (without type definitions), all columns will be set to TEXT (similar to character in R) by default. Setting the right data type for each variable follows essentially the same logic as setting the data types of a data frame's columns in R (with the difference that in SQL this also affects how the data is stored on disk).[12]

In a first step, we thus create a new table called econ.

```
-- Create the new table
CREATE TABLE econ(
"date" DATE,
"pce" REAL,
"pop" REAL,
"psavert" REAL,
"uempmed" REAL,
"unemploy" INTEGER
);
```

Then, we can import the data from the csv file, by first switching to CSV mode via the command .mode csv and then importing the data to econ with .import. The .import command expects as a first argument the path to the CSV file on disk and as a second argument the name of the table to import the data to.

```
-- prepare import
.mode csv
-- import data from csv
.import --skip 1 economics.csv econ
```

Now we can have a look at the new database table in SQLite. .tables shows that we now have one table called econ in our database, and .schema displays the structure of the new econ table.

---

[12]The most commonly used data types in SQL all have a very similar R equivalent: DATE is like Date in R, REAL like numeric/double, INTEGER like integer, and TEXT like character.

```
.tables
```

```
# econ
```

```
.schema econ
```

```
# CREATE TABLE econ(
# "date" DATE,
# "pce" REAL,
# "pop" REAL,
# "psavert" REAL,
# "uempmed" REAL,
# "unemploy" INTEGER
# );
```

With this, we can start querying data with SQLite. In order to make the query results easier to read, we first set two options regarding how query results are displayed on the terminal. `.header on` enables the display of the column names in the
returned query results. And `.mode columns` arranges the query results in columns.

```
.header on
```

```
.mode columns
```

In our first query, we select all (*) variable values of the observation of January 1968.

```
select * from econ where date = '1968-01-01';
```

```
##         date    pce     pop psavert uempmed unemploy
## 1 1968-01-01 531.5 199808    11.7     5.1     2878
```

4.4.1.1 Simple queries

Now let's select all dates and unemployment values of observations with more than
15 million unemployed, ordered by date.

```
select date,
unemploy from econ
where unemploy > 15000
order by date;
```

```
##           date unemploy
## 1 2009-09-01    15009
## 2 2009-10-01    15352
## 3 2009-11-01    15219
## 4 2009-12-01    15098
## 5 2010-01-01    15046
## 6 2010-02-01    15113
## 7 2010-03-01    15202
## 8 2010-04-01    15325
## 9 2010-11-01    15081
```

4.4.2 Joins

So far, we have only considered queries involving one table of data. However, SQL provides a very efficient way to join data from various tables. Again, the way of writing SQL code is the same: You describe what the final table should look like and from where the data is to be selected.

Let's extend the previous example by importing an additional table to our `mydb.sqlite`. The additional data is stored in the file `inflation.csv` in the book's data folder and contains information on the US annual inflation rate measured in percent.[13]

```
-- Create the new table
CREATE TABLE inflation(
"date" DATE,
"inflation_percent" REAL
);


-- prepare import
.mode csv
-- import data from csv
.import --skip 1 inflation.csv inflation
-- switch back to column mode
.mode columns
```

Note that the data stored in `econ` contains monthly observations, while `inflation` contains annual observations. We can thus only meaningfully combine the two datasets at the level of years. Again using the combination of datasets in R as a reference point, here is what we would like to achieve expressed in R. The aim is to

---

[13]Like the data stored in `economics.csv`, the data stored in `inflation.csv` is provided by the Federal Reserve Bank's (FRED)[https://fred.stlouisfed.org/] website.

get a table that serves as basis for a Phillips curve[14] plot, with annual observations and the variables `year`, `average_unemp_percent`, and `inflation_percent`.

```r
# import data
econ <- read.csv("data/economics.csv")
inflation <- read.csv("data/inflation.csv")

# prepare variable to match observations
econ$year <- lubridate::year(econ$date)
inflation$year <- lubridate::year(inflation$date)

# create final output
years <- unique(econ$year)
averages <- sapply(years, FUN = function(x) {
        mean(econ[econ$year==x,"unemploy"]/econ[econ$year==x,"pop"])*100

} )
unemp <- data.frame(year=years,
                    average_unemp_percent=averages)
# combine via the year column
# keep all rows of econ
output<- merge(unemp, inflation[, c("year", "inflation_percent")], by="year")
# inspect output
head(output)
```

```
##   year average_unemp_percent inflation_percent
## 1 1967                 1.512             2.773
## 2 1968                 1.394             4.272
## 3 1969                 1.396             5.462
## 4 1970                 2.013             5.838
## 5 1971                 2.419             4.293
## 6 1972                 2.324             3.272
```

Now let's look at how the same table can be created in SQLite (the table output below only shows the first 6 rows of the resulting table).

```sql
SELECT
strftime('%Y', econ.date)  AS year,
AVG(unemploy/pop)*100 AS average_unemp_percent,
inflation_percent
```

---

[14]https://en.wikipedia.org/wiki/Phillips_curve

```
FROM econ INNER JOIN inflation ON year = strftime('%Y', inflation.date)
GROUP BY year
```

```
##   year average_unemp_percent inflation_percent
## 1 1967                 1.512             2.773
## 2 1968                 1.394             4.272
## 3 1969                 1.396             5.462
## 4 1970                 2.013             5.838
## 5 1971                 2.419             4.293
## 6 1972                 2.324             3.272
```

When done working with the database, we can exit SQLite by typing .quit into the terminal and hit enter.

4.5   With a little help from my friends: GPT and R/SQL coding

Whether you are already an experienced programmer in R and SQL or whether you are rather new to coding, recent developments in Large Language Models (LLMs) might provide an interesting way of making your coding workflow more efficient. At the writing of this book, OpenAI's ChatGPT was still in its testing phase but has already created a big hype in various topic domains. In very simple terms ChatGPT and its predecessors GPT-2, GPT-3 are pre-trained large-scale machine learning models that have been trained on millions of websites' text content (including code from open repositories such as GitHub). Applying these models for predictions is different from other machine learning settings. Instead of feeding new datasets into the trained model, you interact with the model via a prompt (like a chat function). That is, among other things you can pose a question to the model in plain English and get an often very reasonable answer, or you can instruct via the prompt to generate some type of text output for you (given your instructions, and potentially additional input). As the model is trained on natural language texts as well as (documented) computer code, you can ask it to write code for you, for example in SQL or R.

While there are many tools that build on LLMs such as GPT-3 already out there and even more still being developed, I want to explicitly point you to two of those: gptstudio[15], an add-in for Rstudio, providing an easy-to-use interface with some of OpenAI's APIs, and GitHub Copilot[16]. The latter is a professionally developed tool to support your software development workflow by, for example, auto-completing

---

[15]https://github.com/MichelNivard/GPTstudio
[16]https://github.com/features/copilot

the code you are writing. To use GitHub Copilot you need a paid subscription. With a subscription the tool can then be installed as an extension to different code editors (for example Visual Studio Code). However, at the time of writing this book no GitHub Copilot extension for RStudio was available. gptstudio is a much simpler but free alternative to GitHub Copilot and it is explicitly made for RStudio.[17] You will, however, need an OpenAI account and a corresponding OpenAI API key (to get these simply follow the instructions here: https://github.com/MichelNivard/GPTstudio) in order to use the gptstudio-add-in. You will be charged for the queries that gptstudio sends to the OpenAI-API; however there are no fixed costs associated with this setup.

Just to give you an idea of how you could use gptstudio for your coding workflow, consider the following example. After installing the add-in and creating your OpenAI account and API key, you can initiate the chat function of the add-in as follows.

```
# replace "YOUR-API-KEY" with
# your actual key
Sys.setenv(OPENAI_API_KEY = "YOUR-API-KEY")
# open chat window
gptstudio:::chat_gpt_addin()
```

This will cause RStudio to launch a Viewer window. You can pose questions or write instructions to OpenAI's GPT model in the prompt field and send the corresponding query by clicking the "Chat" button. In the example below, I simply ask the model to generate a SQL query for me. In fact, I ask it to construct a query that we have previously built and evaluated in the previous SQL examples. I want the model to specifically reproduce the following query:

```
select date,
unemploy from econ
where unemploy > 15000
order by date;
```

Figure 4.1 shows a screenshot of my instruction to the model, and Figure 4.2 presents the response from the model.

Two things are worth noting here: first, the query is syntactically correct and would essentially work; second, when comparing the query or the query's results with our previous manually written query, we notice that the AI's query is not semantically correct. Our database's unemployment variable is called unemploy, and it is

---

[17]The installation instructions in the README file on https://github.com/MichelNivard/GPTstudio are straightforward.

Write Prompt
The following SQL code returns a table with all dates and unemployment values of observations with more than 15 million unemployed, ordered by date from from a SQL table called "econ"
Chat

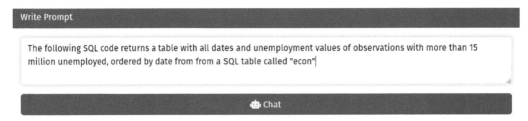

FIGURE 4.1: GPTStudio: instructing OpenAI's GPT-3 model (text-davinci-003) to write an SQL query.

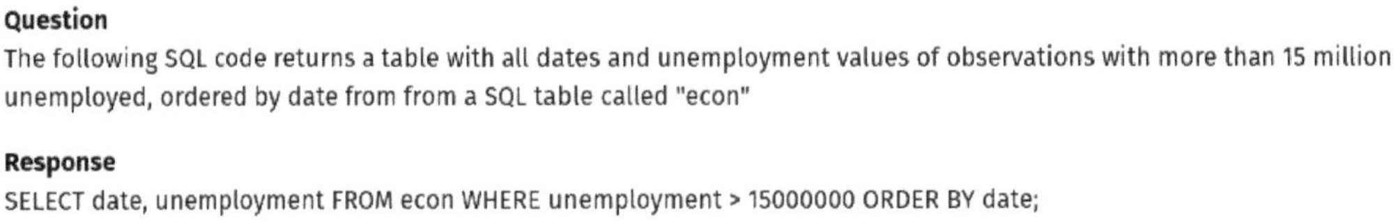

Question
The following SQL code returns a table with all dates and unemployment values of observations with more than 15 million unemployed, ordered by date from from a SQL table called "econ"
Response
SELECT date, unemployment FROM econ WHERE unemployment > 15000000 ORDER BY date;

FIGURE 4.2: GPTStudio: an SQL query written by OpenAI's GPT-3 model (text-davinci-003).

measured in thousands. The GPT model, of course, had no way of obtaining this information from our instructions. As a result, it simply used variable names and values for the filtering that seemed most reasonable given our input. The take-away message here is to be aware of giving the model very clear instructions when creating code in this manner, especially in terms of the broader context (here the database and schema you are working with). To check the model's code for syntax errors, simply test whether the code runs through or not. However, model-generated code can easily introduce semantic errors, which can be very problematic.

4.6 Wrapping up

- Find bottlenecks in your code before exposing it to the full dataset. To do so, use tools like `bench::mark()` and `profvis::profvis()` to see how long certain parts of your code need to process and how much memory they occupy.
- Be aware of R's strengths and weaknesses when writing code for Big Data Analytics.Pre-allocate memory for objects in which you collect the results of loops, make use of R's vectorization, and avoid unnecessary copying.
- Get familiar with SQL and the underlying concept of only loading those observations and variables into R that are really needed for your task. SQLite in combination with R is an excellent lightweight solution to do this.

5

Hardware: Computing Resources

In order to better understand how we can use the available computing resources most efficiently in an analytics task, we first need to get an idea of what we mean by capacity and *big* regarding the most important hardware components. We then look at each of these components (and additional specialized components) through the lens of Big Data. That is, for each component, we look at how it can become a crucial bottleneck when processing large amounts of data and what we can do about it in R. First we focus on mass storage and memory, then on the CPU, and finally on new alternatives to the CPU.

5.1 Mass storage

In a simple computing environment, the mass storage device (hard disk) is where the data to be analyzed is stored. So, in what units do we measure the size of datasets and consequently the mass storage capacity of a computer? The smallest unit of information in computing/digital data is called a *bit* (from *binary digit*; abbrev. 'b') and can take one of two (symbolic) values, either a 0 or a 1 ("off" or "on"). Consider, for example, the decimal number 139. Written in the binary system, 139 corresponds to the binary number 10001011. In order to store this number on a hard disk, we require a capacity of 8 bits, or one *byte* (1 byte = 8 bits; abbrev. 'B'). Historically, one byte encoded a single character of text (e.g., in the ASCII character encoding system). When thinking of a given dataset in its raw/binary representation, we can simply think of it as a row of 0s and 1s.

Bigger units for storage capacity usually build on bytes, for example:

- 1 kilobyte (KB) $= 1000^1 \approx 2^{10}$ bytes
- 1 megabyte (MB) $= 1000^2 \approx 2^{20}$ bytes
- 1 gigabyte (GB) $= 1000^3 \approx 2^{30}$ bytes

Currently, a common laptop or desktop computer has several hundred GBs of mass storage capacity. The problems related to a lack of mass storage capacity in Big Data analytics are likely the easiest to understand. Suppose you collect large amounts of data from an online source such as the Twitter. At some point, R will throw an error and stop the data collection procedure as the operating system will not allow R to use up more disk space. The simplest solution to this problem is to clean up your hard disk: empty the trash, archive files in the cloud or onto an external drive and delete them on the main disk, etc. In addition, there are some easy-to-learn tricks to use from within R to save some disk space.

5.1.1 Avoiding redundancies

Different formats for structuring data stored on disk use up more or less space. A simple example is the comparison of JSON (JavaScript Object Notation) and CSV (Comma Separated Values), both data structures that are widely used to store data for analytics purposes. JSON is much more flexible in that it allows the definition of arbitrarily complex hierarchical data structures (and even allows for hints at data types). However, this flexibility comes with some overhead in the usage of special characters to define the structure. Consider the following JSON excerpt of an economic time series fetched from the Federal Reserve's FRED API[1].

```
{
    "realtime_start": "2013-08-14",
    "realtime_end": "2013-08-14",
    "observation_start": "1776-07-04",
    "observation_end": "9999-12-31",
    "units": "lin",
    "output_type": 1,
    "file_type": "json",
    "order_by": "observation_date",
    "sort_order": "asc",
    "count": 84,
    "offset": 0,
    "limit": 100000,
    "observations": [
        {
            "realtime_start": "2013-08-14",
            "realtime_end": "2013-08-14",
            "date": "1929-01-01",
            "value": "1065.9"
        },
        {
            "realtime_start": "2013-08-14",
            "realtime_end": "2013-08-14",
            "date": "1930-01-01",
```

---

[1]https://fred.stlouisfed.org/docs/api/fred/series_observations.html#example_json

```
            "value": "975.5"
        },
        ...,
        {
            "realtime_start": "2013-08-14",
            "realtime_end": "2013-08-14",
            "date": "2012-01-01",
            "value": "15693.1"
        }
    ]
}
```

The JSON format is very practical here in separating metadata (such as what time frame is covered by this dataset, etc.) in the first few lines on top from the actual data in `"observations"` further down. However, note that due to this structure, the key names like `"date"`, and `"value"` occur for each observation in that time series. In addition, `"realtime_start"` and `"realtime_end"` occur both in the metadata section and again in each observation. Each of those occurrences costs some bytes of storage space on your hard disk but does not add any information once you have parsed and imported the time series into R. The same information could also be stored in a more efficient way on your hard disk by simply storing the metadata in a separate text file and the actual observations in a CSV file (in a table-like structure):

```
"date","value"
"1929-01-01", "1065.9"
"1930-01-01", "975.5"


...,


"2012-01-01", 15693.1"
```

In fact, in this particular example, storing the data in JSON format would take up more than double the hard-disk space as CSV. Of course, this is not to say that one should generally store data in CSV files. In many situations, you might really have to rely on JSON's flexibility to represent more complex structures. However, in practice it is very much worthwhile to think about whether you can improve storage efficiency by simply storing raw data in a different format.

Another related point to storing data in CSV files is to remove redundancies by splitting the data into several tables/CSV files, whereby each table contains the variables exclusively describing the type of observation in it. For example, when analyzing customer data for marketing purposes, the dataset stored in one CSV file might be at the level of individual purchases. That is, each row contains information on what has been purchased on which day by which customer as well as additional variables describing the customer (such as customer ID, name, address, etc.). Instead of keeping all of this data in one file, we could split it into two files, where one only contains the order IDs and corresponding customer IDs as well as attributes of individual orders (but not additional attributes of the customers themselves), and the other contains the customer IDs and all customer attributes. Thereby, we avoid redundancies in the form of repeatedly storing the same values of customer attributes (like name and address) for each order.[2]

5.1.2 Data compression

Data compression essentially follows from the same basic idea of avoiding redundancies in data storage as the simple approaches discussed above. However, it happens on a much more fundamental level. Data compression algorithms encode the information contained in the original representation of the data with fewer bits. In the case of lossless compression, this results in a new data file containing the exact same information but taking up less space on disk. In simple terms, compression replaces repeatedly occurring sequences with shorter expressions and keeps track of replacements in a table. Based on the table, the file can then be de-compressed to recreate the original representation of the data. For example, consider the following character string.

```
"xxxxxyyyyyzzzz"
```

The same data could be represented with fewer bits as:

```
"5x6y4z"
```

which needs fewer than half the number of bits to be stored (but contains the same information).

There are several easy ways to use your mass storage capacity more efficiently with data compression in R. Most conveniently, some functions to import/export data in R directly allow for reading and writing of compressed formats. For example, the `fread()`/`fwrite()` functions provided in the `data.table` package will automatically use the GZIP (de-)compression utility when writing to (reading from) a CSV file with a `.gz` file extension in the file name.

```
# load packages
library(data.table)
```

---

[2]This concept of organizing data into several tables is the basis of relational database management systems, which we will look at in more detail in Chapter 5. However, the basic idea is also very useful for storing raw data efficiently even if there is no intention to later build a database and run SQL queries on it.

```r
# load example data from basic R installation
data("LifeCycleSavings")

# write data to normal csv file and check size
fwrite(LifeCycleSavings, file="lcs.csv")
file.size("lcs.csv")
```

```
## [1] 1441
```

```r
# write data to a GZIPped (compressed) csv file and check size
fwrite(LifeCycleSavings, file="lcs.csv.gz")
file.size("lcs.csv.gz")
```

```
## [1] 744
```

```r
# read/import the compressed data
lcs <- data.table::fread("lcs.csv.gz")
```

Alternatively, you can also use other types of data compression as follows.

```r
# common ZIP compression (independent of data.table package)
write.csv(LifeCycleSavings, file="lcs.csv")
file.size("lcs.csv")
```

```
## [1] 1984
```

```r
zip(zipfile = "lcs.csv.zip", files =  "lcs.csv")
file.size("lcs.csv.zip")
```

```
## [1] 1205
```

```r
# unzip/decompress and read/import data
lcs_path <- unzip("lcs.csv.zip")
lcs <- read.csv(lcs_path)
```

Note that data compression is subject to a time–memory trade-off. Compression and de-compression are computationally intensive and need time. When using compression to make more efficient use of the available mass storage capacity, think about how frequently you expect the data to be loaded into R as part of the data analysis tasks ahead and for how long you will need to keep the data stored on your hard disk. Importing GBs of compressed data can be uncomfortably slower than importing from an uncompressed file.

So far, we have only focused on data size in the context of mass storage capacity. But what happens once you load a large dataset into R (e.g., by means of `read.csv()`)? A program called a "parser" is executed that reads the raw data from the hard disk and creates a representation of that data in the R environment, that is, in random access memory (RAM). All common computers have more GBs of mass storage available than GBs of RAM. Hence, new issues of hardware capacity loom at the stage of data import, which brings us to the next subsection.

5.2   Random access memory (RAM)

Currently, a common laptop or desktop computer has 8–32 GB of RAM capacity. These are more-or-less the numbers you should keep in the back of your mind for the examples/discussions that follow. That is, we will consider a dataset as "big" if it takes up several GBs in RAM (and therefore might overwhelm a machine with 8GB RAM capacity).

There are several types of problems that you might run into in practice when attempting to import and analyze a dataset of the size close to or larger than your computer's RAM capacity. Importing the data might take much longer than expected, your computer might freeze during import (or later during the analysis), R/Rstudio might crash, or you might get an error message hinting at a lack of RAM. How can you anticipate such problems, and what can you do about them?

Many of the techniques and packages discussed in the following chapters are in one way or another solutions to these kinds of problems. However, there are a few relatively simple things to keep in mind before we go into the details.

1. The same data stored on the mass storage device (e.g., in a CSV file) might take up more or less space in RAM. This is due to the fact that the data is (technically speaking) structured differently in a CSV or JSON file than in, for example, a data table or a matrix in R. For example, it is reasonable to anticipate that the example JSON file with the economic time series data will take up less space as a time series object in R (in RAM) than it does on the hard disk (for one thing simply due to the fact that we will not keep the redundancies mentioned before).

2. The import might work well, but some parts of the data analysis script might require much more memory to run through even without loading

additional data from disk. A classic example of this is regression analysis performed with, for example, `lm()` in R. As part of the OLS estimation procedure, `lm` will need to create the model matrix (usually denoted X). Depending on the model you want to estimate, the model matrix might actually be larger than the data frame containing the dataset. In fact, this can happen quite easily if you specify a fixed effects model in which you want to account for the fixed effects via dummy variables (for example, for each country except for one).[3] Again, the result can be one of several: an error message hinting at a lack of memory, a crash, or the computer slowing down significantly. Anticipating these types of problems is very tricky since memory problems are often caused at a lower level of a function from the package that provides you with the data analytics routine you intend to use. Accordingly, error messages can be rather cryptic.

3. Keep in mind that you have some leeway in how much space imported data takes up in R by considering data structures and data types. For example, you can use factors instead of character vectors when importing categorical variables into R (the default in `read.csv`), and for some operations it makes sense to work with matrices instead of data frames.

Finally, recall the lessons regarding memory usage from the section "Writing efficient R code" in Chapter 1.

5.3 Combining RAM and hard disk: Virtual memory

What if all the RAM in our computer is not enough to store all the data we want to analyze?

Modern operating systems (OSs) have a way of dealing with such a situation. Once all RAM is used up by the currently running programs, the OS allocates parts of the memory back to the hard disk, which then works as *virtual memory*. Figure 4.2 illustrates this point.

For example, when we implement an R-script that imports one file after another into the R environment, ignoring the RAM capacity of our computer, the OS will start *paging* data to the virtual memory. This happens 'under the hood' without explicit instructions by the user. We will quite likely notice that the computer slows down a lot when this happens.

---

[3] For example, if you specify something like `lm(y~x1 + x2 + country, data=mydata)` and `country` is a categorical variable (factor).

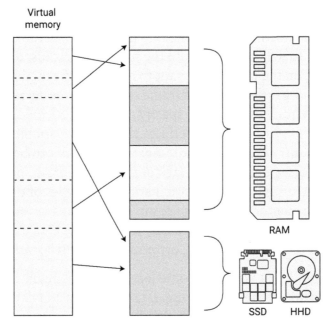

FIGURE 5.1: Virtual memory. Overall memory is mapped to RAM and parts of the hard disk.

While this default usage of virtual memory by the OS is helpful for running several applications at the same time, each taking up a moderate amount of memory, it is not a really useful tool for processing large amounts of data in one application (R). However, the underlying idea of using both RAM and mass storage simultaneously in order to cope with a lack of memory is very useful in the context of Big Data Analytics.

Several R packages have been developed that exploit the idea behind virtual memory explicitly for analyzing large amounts of data. The basic idea behind these packages is to map a dataset to the hard disk when loading it into R. The actual data values are stored in chunks on the hard disk, while the structure/metadata of the dataset is loaded into R.

5.4   CPU and parallelization

The actual processing of the data is done in the computer's central processing unit (CPU). Consequently, the performance of the CPU has a substantial effect on how fast a data analytics task runs. A CPU's performance is usually denoted by its *clock rate* measured in gigaherz (GHz). In simple terms, a CPU with a clock rate of 4.8 GHz can execute 4.8 billion basic operations per second. Holding all other aspects constant, you can thus expect an analytics task to run faster if it runs on a computer with higher CPU clock rate. As an alternative to scaling up the CPU, we can exploit the fact that modern CPUs have several *cores*. In the normal usage of a PC, the operating system makes use of these cores to run several applications smoothly *in parallel* (e.g., you listen to music on Spotify while browsing the web and running some analytics script in RStudio in the background).

Modern computing environments such as R allow us to explicitly run parts of the same analytics task in parallel, that is, on several CPU cores at the same time. Following the same logic, we can also connect several computers (each with several CPU cores) in a cluster computer and run the program in parallel on all of these computing nodes. Both of these approaches are generally referred to as *parallelization*, and both are supported in several R packages.

An R program run in parallel typically involves the following steps.

- First, several instances of R are running at the same time (across one machine with multiple CPU cores or across a cluster computer). One of the instances (i.e., the *master* instance) breaks the computation into batches and sends those to the other instances.
- Second, each of the instances processes its batch and sends the results back to the master instance.
- Finally, the master instance combines the partial results into the final result and returns it to the user.

To illustrate this point, consider the following econometric problem: you have a customer dataset[4] with detailed data on customer characteristics, past customer behavior, and information on online marketing campaigns. Your task is to figure out which customers are more likely to react positively to the most recent online marketing campaign. The aim is to optimize personalized marketing campaigns in the future based on insights gained from this exercise. In a first step you take a computationally intensive "brute force" approach: you run all possible regressions with the dependent variable Response (equal to 1 if the customer took the offer in the campaign and 0 otherwise). In total you have 21 independent variables; thus you need to run $2^{20} = 1,048,576$ logit regressions (this is without considering linear combinations of covariates etc.). Finally, you want to select the model with the best fit according to deviance.

A simple sequential implementation to solve this problem could look like this (for the sake of time, we cap the number of regression models to N=10).

---

[4]https://www.kaggle.com/jackdaoud/marketing-data?select=marketing_data.csv

```r
# you can download the dataset from
# https://www.kaggle.com/jackdaoud/marketing-data?
# select=marketing_data.csv

# PREPARATION ---------------------------
# packages
library(stringr)

# import data
marketing <- read.csv("data/marketing_data.csv")
# clean/prepare data
marketing$Income <- as.numeric(gsub("[[:punct:]]",
                                    "",
                                    marketing$Income))
marketing$days_customer <-
    as.Date(Sys.Date())-
    as.Date(marketing$Dt_Customer, "%m/%d/%y")
marketing$Dt_Customer <- NULL

# all sets of independent vars
indep <- names(marketing)[ c(2:19, 27,28)]
combinations_list <- lapply(1:length(indep),
                            function(x) combn(indep, x,
                                              simplify = FALSE))
combinations_list <- unlist(combinations_list,
                            recursive = FALSE)
models <- lapply(combinations_list,
                 function(x) paste("Response ~",
                                   paste(x, collapse="+")))

# COMPUTE REGRESSIONS -------------------------
N <- 10 #  N <- length(models) for all
pseudo_Rsq <- list()
length(pseudo_Rsq) <- N
for (i in 1:N) {
  # fit the logit model via maximum likelihood
  fit <- glm(models[[i]],
             data=marketing,
             family = binomial())
  # compute the proportion of deviance explained by
  # the independent vars (~R^2)
```

```
  pseudo_Rsq[[i]] <- 1-(fit$deviance/fit$null.deviance)
}


# SELECT THE WINNER --------------
models[[which.max(pseudo_Rsq)]]


## [1] "Response ~ MntWines"
```

Alternatively, a sequential implementation could be based on an apply-type function like `lapply()`. As several of the approaches to parallelize computation with R build either on loops or an apply-type syntax, let us also briefly introduce the sequential lapply-implementation of the task above as a point of reference.

```
# COMPUTE REGRESSIONS -------------------------
N <- 10 #  N <- length(models) for all
run_reg <-
    function(model, data, family){
        # fit the logit model via maximum likelihood
        fit <- glm(model, data=data, family = family)
        # compute and return the proportion of deviance explained by
        # the independent vars (~R^2)
        return(1-(fit$deviance/fit$null.deviance))
    }

pseudo_Rsq_list <-lapply(models[1:N], run_reg, data=marketing, family=binomial() )
pseudo_Rsq <- unlist(pseudo_Rsq_list)

# SELECT THE WINNER --------------
models[[which.max(pseudo_Rsq)]]


## [1] "Response ~ MntWines"
```

5.4.1   Naive multi-session approach

There is actually a simple way of doing this "manually" on a multi-core PC, which intuitively illustrates the point of parallelization (although it would not be a very practical approach): you write an R script that loads the dataset, runs the first n of the total of N regressions, and stores the result in a local text file. Next, you run the script in your current RStudio session, open an additional RStudio session, and run the script with the next n regressions, and so on until all cores are occupied with one RStudio session. At the end you collect all of the results from the separate text files and combine them to get the final result. Depending on the problem at hand, this could indeed speed up the overall task, and it is technically speaking a form of "multi-session" approach. However, as you have surely noticed, this is unlikely to be a very practical approach.

5.4.2 Multi-session approach with futures

There is a straightforward way to implement the very basic (naive) idea of running parts of the task in separate R sessions. The `future` package (see Bengtsson (2021) for details) provides a lightweight interface (API) to use futures[5]. An additional set of packages (such as `future.apply`) that build on the `future` package, provides high-level functionality to run your code in parallel without having to change your (sequential, usual) R code much. In order to demonstrate the simplicity of this approach, let us re-write the sequential implementation through `lapply()` from above for parallelization through the `future` package. All we need to do is to load the `future` and `future.apply` packages (Bengtsson, 2021) and then simply replace `lapply(...)` with `future_lapply(...)`.

```
# SET UP ------------------

# load packages
library(future)
library(future.apply)
# instruct the package to resolve
# futures in parallel (via a SOCK cluster)
plan(multisession)


# COMPUTE REGRESSIONS -------------------------
N <- 10 #  N <- length(models) for all
pseudo_Rsq_list <- future_lapply(models[1:N],
                          run_reg,
                          data=marketing,
                          family=binomial() )
pseudo_Rsq <- unlist(pseudo_Rsq_list)


# SELECT THE WINNER --------------
models[[which.max(pseudo_Rsq)]]


## [1] "Response ~ MntWines"
```

---

[5]In simple terms, futures are a programming concept that allows for the asynchronous execution of a task. A future is a placeholder object that represents the result of an asynchronous operation. The future object can be used to check the status of the asynchronous operation and to retrieve the result of the operation when it is completed. By using futures, tasks can be broken down into smaller tasks that can be executed in parallel, resulting in faster completion times.

5.4.3 Multi-core and multi-node approach

There are several additional approaches to parallelization in R. With the help of some specialized packages, we can instruct R to automatically distribute the workload to different cores (or different computing nodes in a cluster computer), control and monitor the progress in all cores, and then automatically collect and combine the results from all cores. The future-package and the packages building on it provide in themselves different approaches to writing such scripts.[6] Below, we look at two additional ways of implementing parallelization with R that are based on other underlying frameworks than future.

5.4.3.1 Parallel for-loops using socket

Probably the most intuitive approach to parallelizing a task in R is the foreach package (Microsoft and Weston, 2022). It allows you to write a foreach statement that is very similar to the for-loop syntax in R. Hence, you can straightforwardly "translate" an already implemented sequential approach with a common for-loop to a parallel implementation.

```
# COMPUTE REGRESSIONS IN PARALLEL (MULTI-CORE) --------------------------

# packages for parallel processing
library(parallel)
library(doSNOW)

# get the number of cores available
ncores <- parallel::detectCores()
# set cores for parallel processing
ctemp <- makeCluster(ncores)
registerDoSNOW(ctemp)

# prepare loop
N <- 10000 #  N <- length(models) for all
# run loop in parallel
pseudo_Rsq <-
  foreach ( i = 1:N, .combine = c) %dopar% {
    # fit the logit model via maximum likelihood
    fit <- glm(models[[i]],
               data=marketing,
               family = binomial())
```

---

[6]See https://www.futureverse.org/packages-overview.html for an overview of the *futureverse*, and https://www.futureverse.org/#ref-bengtsson-future for a set of simple examples of using future in different ways (with different syntaxes/coding styles).

```
    # compute the proportion of deviance explained by
    # the independent vars (~R^2)
    return(1-(fit$deviance/fit$null.deviance))
}


# SELECT THE WINNER --------------
models[[which.max(pseudo_Rsq)]]


## [1] "Response ~ Year_Birth+Teenhome+Recency+MntWines+days_customer"
```

With relatively few cases, this approach is not very fast due to the overhead of "distributing" variables/objects from the master process to all cores/workers. In simple terms, the socket approach means that the cores do not share the same variables/the same environment, which creates overhead. However, this approach is usually very stable and runs on all platforms.

5.4.3.2   Parallel lapply using forking

Finally, let us look at an implementation based on forking (here, implemented in the `parallel` package by (R Core Team, 2021). In the fork approach, each core works with the same objects/variables in a shared environment, which makes this approach very fast. However, depending on what exactly is being computed, sharing an environment can cause problems.[7] If you are not sure whether your setup might run into issues with forking, it would be better to rely on a non-fork approach.[8]

```
# COMPUTE REGRESSIONS IN PARALLEL (MULTI-CORE) --------------


# prepare parallel lapply (based on forking,
# here clearly faster than foreach)
N <- 10000 #  N <- length(models) for all
# run parallel lapply
pseudo_Rsq <- mclapply(1:N,
                      mc.cores = ncores,
                      FUN = function(i){
                        # fit the logit model
                        fit <- glm(models[[i]],
                                    data=marketing,
```

---

[7] Also, this approach does not work on Windows machines (see `?mclapply` for details).

[8] Note that you can use the forking approach also to resolve futures. By setting `plan(multicore)` instead of `plan(multisession)` when working with the `future` package, parallelization based on futures will be run via forking (again, this will not work on Windows machines).

```
                                  family = binomial())
              # compute the proportion of deviance
              # explained  by the independent vars (~R^2)
              return(1-(fit$deviance/fit$null.deviance))
              })


# SELECT THE WINNER, SHOW FINAL OUTPUT --------------


best_model <- models[[which.max(pseudo_Rsq)]]
best_model


## [1] "Response ~ Year_Birth+Teenhome+Recency+MntWines+days_customer"
```

5.5 GPUs for scientific computing

The success of the computer games industry in the late 1990s/early 2000s led to an interesting positive externality for scientific computing. The ever more demanding graphics of modern computer games and the huge economic success of the computer games industry set incentives for hardware producers to invest in research and development of more powerful 'graphics cards', extending a normal PC/computing environment with additional computing power solely dedicated to graphics. At the heart of these graphic cards are so-called GPUs (graphics processing units), microprocessors specifically optimized for graphics processing. Figure 5.2 depicts a modern graphics card similar to those commonly built into today's 'gaming' PCs.

FIGURE 5.2: Illustration of a Nvidia GEFORCE RTX 2080 graphics card with a modern GPU (illustration by MarcusBurns1977 under CC BY 3.0 license).

Why did the hardware industry not simply invest in the development of more powerful CPUs to deal with the more demanding PC games? The main reason is that the architecture of CPUs is designed not only for efficiency but also flexibility. That is, a CPU needs to perform well in all kinds of computations, some parallel, some sequential, etc. Computing graphics is a comparatively narrow domain of computation, and designing a processing unit architecture that is custom-made to excel just at this one task is thus much more cost efficient. Interestingly, this graphics-specific architecture (specialized in highly parallel numerical [floating point] workloads) turns out to also be very useful in some core scientific computing tasks – in particular, matrix multiplications (see Fatahalian et al. (2004) for a detailed discussion of why that is the case). A key aspect of GPUs is that they are composed of several multiprocessor units, of which each in turn has several cores. GPUs can thus perform computations with hundreds or even thousands of threads in parallel. The figure below illustrates this point by showing the typical architecture of an NVIDIA GPU.

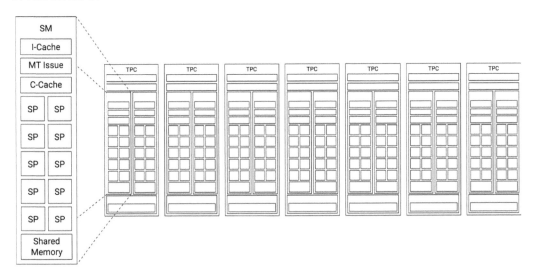

FIGURE 5.3: Illustration of a graphics processing unit's components/architecture. The GPU consists of several Texture Processing Clusters (TPC), which in turn consist of several Streaming Multiprocessors (SM; the primary unit of parallelism in the GPU) that contain ten Streaming Processors (SP; cores, responsible for executing a single thread), shared memory (can be accessed by multiple SPs simultaneously), instruction cache (I-Cache; responsible for storing and managing the instructions needed to execute a program), constant cache (C-Cache; store constant data that is needed during program execution), and a multi-threaded issue component (MT issue; responsible for scheduling and managing the execution of multiple threads simultaneously).

While initially, programming GPUs for scientific computing required a very good understanding of the hardware, graphics card producers have realized that there is an additional market for their products (in particular with the recent rise of deep learning) and now provide several high-level APIs to use GPUs for tasks other than graphics processing. Over the last few years, more high-level software has been developed that makes it much easier to use GPUs in parallel computing tasks. The following subsections show some examples of such software in the R environment.[9]

5.5.1 GPUs in R

The gpuR package (Determan, 2019) provides basic R functions to compute with GPUs from within the R environment.[10] In the following example we compare the performance of a CPU with a GPU for a matrix multiplication exercise. For a large $N \times P$ matrix X, we want to compute $X^t X$.

In a first step, we load the gpuR package.[11] Note the output to the console. It shows the type of GPU identified by gpuR. This is the platform on which gpuR will compute the GPU examples. In order to compare the performances, we also load the bench package.

```
# load package
library(bench)
library(gpuR)
```

```
## Number of platforms: 1
## - platform: NVIDIA Corporation: OpenCL 3.0 CUDA 12.0.151
##    - context device index: 0
##       - NVIDIA GeForce GTX 1650
## checked all devices
## completed initialization
```

Note how loading the gpuR package triggers a check of GPU devices and outputs information on the detected GPUs as well as the lower-level software platform to run GPU computations. Next, we initialize a large matrix filled with pseudo-random numbers, representing a dataset with N observations and P variables.

---

[9]Note that while these examples are easy to implement and run, setting up a GPU for scientific computing can still involve many steps and some knowledge of your computer's system. The examples presuppose that all installation and configuration steps (GPU drivers, CUDA, etc.) have already been completed successfully.

[10]See https://github.com/cdeterman/gpuR/wiki for installation instructions regarding the dependencies. Once the package dependencies are installed you can install the gpuR-package directly from GitHub: devtools::install_github("cdeterman/gpuR") (make sure the devtools-package is installed before doing this).

[11]As with setting up GPUs on your machine in general, installing all prerequisites to make gpuR work on your local machine can be a bit of work and can depend a lot on your system.

```
# initialize dataset with pseudo-random numbers
N <- 10000  # number of observations
P <- 100 # number of variables
X <- matrix(rnorm(N * P, 0, 1), nrow = N, ncol =P)
```

For the GPU examples to work, we need one more preparatory step. GPUs have their own memory, which they can access faster than they can access RAM. However, this GPU memory is typically not very large compared to the memory CPUs have access to. Hence, there is a potential trade-off between losing some efficiency but working with more data or vice versa.[12] Here, we transfer the matrix to GPU memory with vclMatrix().[13].

```
# prepare GPU-specific objects/settings
# transfer matrix to GPU (matrix stored in GPU memory)
vclX <- vclMatrix(X, type = "float")
```

Now we run the two examples, first, based on standard R, using the CPU, and then, computing on the GPU and using GPU memory. In order to make the comparison fair, we force bench::mark() to run at least 200 iterations per variant.

```
# compare three approaches
gpu_cpu <- bench::mark(

  # compute with CPU
  cpu <-t(X) %*% X,

  # GPU version, in GPU memory
  # (vclMatrix formation is a memory transfer)
  gpu <- t(vclX) %*% vclX,

check = FALSE, memory = FALSE, min_iterations = 200)
```

The performance comparison is visualized with boxplots.

---

[12] If we instruct the GPU to use its own memory but the data does not fit in it, the program will result in an error.

[13] Alternatively, with gpuMatrix() we can create an object representing matrix x for computation on the GPU, while only pointing the GPU to the matrix and without actually transferring data to the GPU's memory.

```
plot(gpu_cpu, type = "boxplot")
```

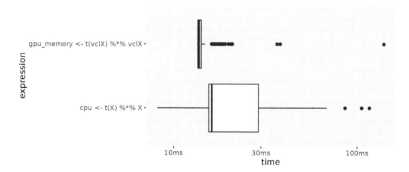

The theoretically expected pattern becomes clearly visible. When using the GPU + GPU memory, the matrix operation is substantially faster than the common CPU computation. However, in this simple example of only one matrix operation, the real strength of GPU computation vs. CPU computation does not really become visible. In Chapter 13, we will look at a computationally much more intensive application of GPUs in the domain of deep learning (which relies heavily on matrix multiplications).

5.6 The road ahead: Hardware made for machine learning

Due to the high demand for more computational power in the domain of training complex neural network models (for example, in computer vision), Google has recently developed a new hardware platform specifically designed to work with complex neural networks using TensorFlow: Tensor Processing Units (TPUs). TPUs were designed from the ground up to improve performance in dense vector and matrix computations with the aim of substantially increasing the speed of training deep learning models implemented with TensorFlow (Abadi et al., 2015).

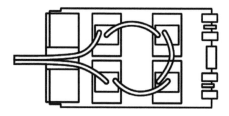

FIGURE 5.4: Illustration of a tensor processing unit (TPU).

While initially only used internally by Google, the Google Cloud platform now offers cloud TPUs to the general public.

5.7   Wrapping up

- Be aware of and avoid redundancies in data storage. Consider, for example, storing data in CSV-files instead of JSON-files (if there is no need to store hierarchical structures).
- File compression is a more general strategy to avoid redundancies and save mass storage space. It can help you store even large datasets on disk. However, reading and saving compressed files takes longer, as additional processing is necessary. As a rule of thumb, store the raw datasets (which don't have to be accessed that often) in a compressed format.
- Standard R for data analytics expects the datasets to be imported and available as R objects in the R environment (i.e., in RAM). Hence, the step of importing large datasets to R with the conventional approaches is aimed to parsing and loading the entire dataset into RAM, which might fail if your dataset is larger than the available RAM.
- Even if a dataset is not too large to fit into RAM, running data analysis scripts on it might then lead to R reaching RAM limits due to the creation of additional R objects in RAM needed for the computation. For example, when running regressions in the conventional way in R, R will generate, among other objects, an object containing the model matrix. However, at this point your original dataset object will still also reside in RAM. Not uncommonly, R would then crash or slow down substantially.
- The reason R might slow down substantially when working with large datasets in RAM is that your computer's operating system (OS) has a default approach of handling situations with a lack of available RAM: it triggers *paging* between the RAM and a dedicated part of the hard disk called *virtual memory*. In simple terms, your computer starts using parts of the hard disk as an extension of RAM. However, reading/writing from/to the hard disk is much slower than from/to RAM, so your entire data analytics script (and any other programs running at the same time) will slow down substantially.
- Based on the points above, when working locally with a large dataset, recognize why your computer is slowing down or why R is crashing. Consider whether the dataset could theoretically fit into memory. Clarify whether analyzing the already imported data triggers the OS's virtual memory mechanism.
- Taken together, your program might run slower than expected due to a lack of RAM (and thus the paging) and/or due to a very high computational burden on the CPU – for example, bootstrapping the standard errors of regression coefficients.
- By default essentially all basic R functions use one CPU thread/core for computation. If RAM is not an issue, setting up repetitive tasks to run in parallel (i.e.,

using more than one CPU thread/core at a time) can substantially speed up your program. Easy-to-use solutions to do this are `foreach` for a parallel version of `for`-loops and `mclapply` for a parallel version of `lapply`.

- Finally, if your analytics script builds extensively on matrix multiplication, consider implementing it for processing on your GPU via the `gpuR` package. Note, though, that this approach presupposes that you have installed and set up your GPU with the right drivers to use it not only for graphics but also for scientific computation.

5.8   Still have insufficient computing resources?

When working with very large datasets (i.e., terabytes of data), processing the data on one common computer might not work due to a lack of memory or would be way too slow due to a lack of computing power (CPU cores). The architecture or basic hardware setup of a common computer is subject to a limited amount of RAM and a limited number of CPUs/CPU cores. Hence, simply scaling up might not be sufficient. Instead, we need to scale out. In simple terms, this means connecting several computers (each with its own RAM, CPU, and mass storage) in a network, distributing the dataset across all computers ("nodes") in this network, and working on the data simultaneously across all nodes. In the next chapter, we look into how such "distributed systems'' basically work, what software frameworks are commonly used to work on distributed systems, and how we can interact with this software (and the distributed system) via R and SQL.

6

Distributed Systems

When we connect several computers in a network to jointly process large amounts of data, such a computing system is commonly referred to as a "distributed system". From a technical standpoint the key difference between a distributed system and the more familiar parallel system (e.g., our desktop computer with its multi core CPU) is that in distributed systems the different components do not share the same memory (and storage). Figure 6.1 illustrates this point.

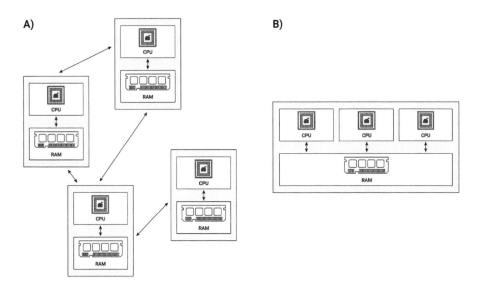

FIGURE 6.1: Panel A illustrates a distributed system, in contrast to the illustration of a parallel system in Panel B.

In a distributed system, the dataset is literally split up into pieces that then reside separately on different nodes. This requires an additional layer of software (that coordinates the distribution/loading of data as well as the simultaneous processing) and different approaches (different programming models) to defining computing/data analytics tasks. Below, we will look at each of these aspects in turn.

6.1 MapReduce

A broadly used programming model for processing Big Data on distributed systems is called MapReduce. It essentially consists of two procedures and is conceptually very close to the "split-apply-combine" strategy in data analysis. First, the Map function sorts/filters the data (on each node/computer). Then, a Reduce function aggregates the sorted/filtered data. Thereby, all of these processes are orchestrated to run across many nodes of a cluster computer. Finally, the master node collects the results and returns them to the user.

Let us illustrate the basic idea behind MapReduce with a simple example. Suppose you are working on a text mining task in which all the raw text in thousands of digitized books (stored as text files) need to be processed. In a first step, you want to compute word frequencies (count the number of occurrences of specific words in all books combined).

For simplicity, let us only focus on the following very simple and often referred to MapReduce word count example[1]:

Text in book 1:

Apple Orange Mango

Orange Grapes Plum

Text in book 2:

Apple Plum Mango

Apple Apple Plum

The MapReduce procedure is then as follows:

- First, the data is loaded from the original text files.
- Each line of text is then passed to individual mapper instances, which separately split the lines of text into key–value pairs. In the example above, the first key–value pair of the first document/line would then be *Apple,1*.
- Then the system sorts and shuffles all key–value pairs across all instances; next, the reducer aggregates the sorted/shuffled key–value pairs (here: counts the number of word occurrences). In the example above, this means all values with key *Apple* are summed up, resulting in *Apple,4*.
- Finally, the master instance collects all the results and returns the final output.

---

[1]See, e.g., https://commons.wikimedia.org/wiki/File:WordCountFlow.JPG for an illustration of the same example.

The result would be as follows:

Text in book 2:

Apple,4

Grapes,1

Mango,2

Orange,2

Plum,3

From this simple example, a key aspect of MapReduce should become clear: for the key tasks of mapping and reducing, the data processing on one node/instance can happen completely independently of the processing on the other instances. Note that this is not as easily achievable for every data analytics task as it is for computing word frequencies.

Aside: MapReduce concept illustrated in R

In order to better understand the basic concept behind the MapReduce framework on a distributed system, let's look at how we can combine the functions `map()` and `reduce()` in R to implement the basic MapReduce example shown above (this is just to illustrate the underlying idea, *not* to suggest that MapReduce actually is simply an application of the classical map and reduce (fold) functions in functional programming).[a] The overall aim of the program is to count the number of times each word is repeated in a given text. The input to the program is thus a text, and the output is a list of key–value pairs with the unique words occurring in the text as keys and their respective number of occurrences as values.

In the code example, we will use the following text as input.

```
# initialize the input text (for simplicity as one text string)
input_text <-
"Apple Orange Mango
Orange Grapes Plum
Apple Plum Mango
Apple Apple Plum"
```

Mapper

The Mapper first splits the text into lines and then splits the lines into key–value pairs, assigning to each key the value 1. For the first step we use `strsplit()`, which takes a character string as input and splits it into a list of sub-strings according to the matches of a sub-string (here "\n", indicating the end of a line).

```
# Mapper splits input into lines
lines <- as.list(strsplit(input_text, "\n")[[1]])
lines[1:2]
```

```
## [[1]]
## [1] "Apple Orange Mango"
##
## [[2]]
## [1] "Orange Grapes Plum"
```

In a second step, we apply our own function (map_fun()) to each line of text via Map(). map_fun() splits each line into words (keys) and assigns a value of 1 to each key.

```
# Mapper splits lines into key-value pairs
map_fun <-
    function(x){

        # remove special characters
        x_clean <- gsub("[[:punct:]]", "", x)
        # split line into words
        keys <- unlist(strsplit(x_clean, " "))
        # initialize key-value pairs
        key_values <- rep(1, length(keys))
        names(key_values) <- keys

        return(key_values)

    }

kv_pairs <- Map(map_fun, lines)

# look at the result
kv_pairs[1:2]
```

```
## [[1]]
##  Apple Orange  Mango
##      1      1      1
##
## [[2]]
```

```
## Orange Grapes     Plum
##       1       1       1
```

Reducer

The Reducer first sorts and shuffles the input from the Mapper and then reduces the key–value pairs by summing up the values for each key.

```
# order and shuffle
kv_pairs <- unlist(kv_pairs)
keys <- unique(names(kv_pairs))
keys <- keys[order(keys)]
shuffled <- lapply(keys,
                   function(x) kv_pairs[x == names(kv_pairs)])
shuffled[1:2]
```

```
## [[1]]
## Apple Apple Apple Apple
##     1     1     1     1
##
## [[2]]
## Grapes
##      1
```

Now we can sum up the keys to get the word count for the entire input.

```
sums <- lapply(shuffled, Reduce, f=sum)
names(sums) <- keys
sums[1:2]
```

```
## $Apple
## [1] 4
##
## $Grapes
## [1] 1
```

---

[a]For a more detailed discussion of what map and reduce *actually* have to do with MapReduce, see https://medium.com/@jkff/mapreduce-is-not-functional-programming-39109a4ba7b2.

6.2   Apache Hadoop

Hadoop MapReduce is the most widely known and used implementation of the MapReduce framework. A decade ago, Big Data Analytics with really large datasets often involved directly interacting with/working in Hadoop to run MapReduce jobs. However, over the last few years various higher-level interfaces have been developed that make the usage of MapReduce/Hadoop by data analysts much more easily accessible. The purpose of this section is thus to give a lightweight introduction to the underlying basics that power some of the code examples and tutorials discussed in the data analytics chapters toward the end of this book.

6.2.1   Hadoop word count example

To get an idea of what running a Hadoop job looks like, we run the same simple word count example introduced above on a local Hadoop installation. The example presupposes a local installation of Hadoop version 2.10.1 (see Appendix C for details) and can easily be run on a completely normal desktop/laptop computer running Ubuntu Linux. As a side remark, this actually illustrates an important aspect of developing MapReducescripts in Hadoop (and many of the software packages building on it): the code can easily be developed and tested locally on a small machine and only later transferred to the actual Hadoop cluster to be run on the full dataset.

The basic Hadoop installation comes with a few templates for very typical map/reduce programs.[2] Below we replicate the same word-count example as shown in simple R code above.

In a first step, we create an input directory where we store the input file(s) to feed to Hadoop.

```
# create directory for input files (typically text files)
mkdir ~/input
```

Then we add a text file containing the same text as in the example above.

```
echo "Apple Orange Mango
Orange Grapes Plum
Apple Plum Mango
Apple Apple Plum" >>  ~/input/text.txt
```

---

[2]More sophisticated programs need to be custom made, written in Java.

Now we can run the MapReduce/Hadoop word count as follows, storing the results in a new directory called `wordcount_example`. We use the already-implemented Hadoop script to run a word count job, MapReduce style. This is where we rely on the already implemented word-count example provided with the Hadoop installation (located in `/usr/local/hadoop/share/hadoop/mapreduce/hadoop-mapreduce-examples-2.10.1.jar`).

```
# run mapreduce word count
/usr/local/hadoop/bin/hadoop jar \
/usr/local/hadoop/share/hadoop/mapreduce/hadoop-mapreduce-examples-2.10.1.jar \
wordcount
~/input ~/wc_example
```

What this line says is: Run the Hadoop program called `wordcount` implemented in the jar-file `hadoop-mapreduce-examples-2.10.1.jar`; use the files in directory `~/input` containing the raw text as input, and store the final output in directory `~/wc_example`.

```
cat ~/wc_example/*
```

```
## Apple    4
## Grapes   1
## Mango    2
## Orange   2
## Plum 3
```

What looks rather simple in this example can get very complex once you want to write an entire data analysis script with all kinds of analysis for Hadoop. Also, Hadoop was designed for batch processing and does not offer a simple interface for interactive sessions. All of this makes it rather impractical for a typical analytics workflow as we know it from working with R. This is where Apache Spark[3] (Zaharia et al., 2016) comes to the rescue.

6.3 Apache Spark

Spark (Zaharia et al., 2016) is a data analytics engine specifically designed for processing large amounts of data on cluster computers. It partially builds on the broader Apache Hadoop framework for handling storage and resource

---

[3]https://spark.apache.org/

management, but it is often faster than Hadoop MapReduce by an order of magnitude. In addition, it offers many more easy-to-use high-level interfaces for typical analytics tasks than Hadoop. In contrast to Hadoop, Spark is specifically made for interactively developing and running data analytics scripts and is therefore more easily accessible to people with an applied econometrics background but no substantial knowledge in MapReduce and/or cluster computing. In particular, it comes with several high-level operators that make it rather easy to implement analytics tasks. As we will see in later chapters, it is very easy to use interactively from within R (and other languages like Python, SQL, and Scala). This makes the platform much more accessible and worthwhile for empirical economic research, even for relatively simple econometric analyses.

The following figure illustrates the basic components of Spark. The main functionality includes memory management, task scheduling, and the implementation of Spark's capabilities to handle and manipulate data distributed across many nodes in parallel. Several built-in libraries extend the core implementation, covering specific domains of practical data analytics tasks (querying structured data via SQL, processing streams of data, machine learning, and network/graph analysis). The last two provide various common functions/algorithms frequently used in data analytics/applied econometrics, such as generalized linear regression, summary statistics, and principal component analysis.

At the heart of Big Data Analytics with Spark is the fundamental data structure called 'resilient distributed dataset' (RDD). When loading/importing data into Spark, the data is automatically distributed across the cluster in RDDs (~ as distributed collections of elements), and manipulations are then executed in parallel on these RDDs. However, the entire Spark framework also works locally on a simple laptop or desktop computer. This is a great advantage when learning Spark and when testing/debugging an analytics script on a small sample of the real dataset.

6.4   Spark with R

There are two prominent packages for using Spark in connection with R: SparkR (Venkataraman et al., 2021) and RStudio's sparklyr (Luraschi et al., 2022). The former is in some ways closer to Spark's Python API; the latter is closer to the dplyr-type of data handling (and is compatible with the tidyverse (Wickham et al., 2019)).[4] For the very simple introductory examples below, either package could

---

[4]See https://cosminsanda.com/posts/a-compelling-case-for-sparkr/ for a more detailed comparison and discussion of advantages of either package.

have been used equally well. For the general introduction we focus on SparkR and later have a look at a simple regression example based on sparklyr.

To install and use Spark from the R shell, only a few preparatory steps are needed. The following examples are based on installing/running Spark on a Linux machine with the SparkR package. SparkR depends on Java (version 8). Thus, we first should make sure the right Java version is installed. If several Java versions are installed, we might have to select version 8 manually via the following terminal command (Linux):

```
# might have to switch to java version 8 first
sudo update-alternatives --config java
```

With the right version of Java running, we can install SparkR from GitHub (needs the devtools package (Wickham et al., 2022)) devtools::install_github("cran/SparkR"). After installing SparkR, the call SparkR::install.spark() will download and install Apache Spark to a local directory.[5] Now we can start an interactive SparkR session from the terminal with

```
$ SPARK-HOME/bin/sparkR
```

where SPARK-HOME is a placeholder for the path to your local Spark installation (printed to the console after running SparkR::install.spark()). Or simply run SparkR from within RStudio by loading SparkR and initiating Spark with sparkR.session().

```
# to install use
# devtools::install_github("cran/SparkR")
# load packages
library(SparkR)
# start session
sparkR.session()
```

By default this starts a local stand-alone session (no connection to a cluster computer needed). While the examples below are all intended to run on a local machine,

---

[5]Note that after the installation, the location of Spark is printed to the R console. Alternatively, you can also first install the sparklyr package and then run sparklyr::spark_install() to install Spark. In the data analysis examples later in the book, we will work both with SparkR and sparklyr.

it is straightforward to connect to a remote Spark cluster and run the same examples there.[6]

6.4.1   Data import and summary statistics

First, we want to have a brief look at how to perform the first few steps of a typical econometric analysis: import data and compute summary statistics. We will analyze the already familiar `flights.csv` dataset. The basic Spark installation provides direct support to import common data formats such as CSV and JSON via the `read.df()` function (for many additional formats, specific Spark libraries are available). To import `flights.csv`, we set the `source` argument to `"csv"`.

```
# Import data and create a SparkDataFrame
# (a distributed collection of data, RDD)
flights <- read.df("data/flights.csv", source = "csv", header="true")

# inspect the object
class(flights)
```

```
## [1] "SparkDataFrame"
## attr(,"package")
## [1] "SparkR"
```

```
dim(flights)
```

```
## [1] 336776      19
```

By default, all variables have been imported as type `character`. For several variables this is, of course, not the optimal data type to compute summary statistics. We thus first have to convert some columns to other data types with the `cast` function.

```
flights$dep_delay <- cast(flights$dep_delay, "double")
flights$dep_time <- cast(flights$dep_time, "double")
flights$arr_time <- cast(flights$arr_time, "double")
flights$arr_delay <- cast(flights$arr_delay, "double")
flights$air_time <- cast(flights$air_time, "double")
flights$distance <- cast(flights$distance, "double")
```

---

[6]Simply set the `master` argument of `sparkR.session()` to the URL of the Spark master node of the remote cluster. Importantly, the local Spark and Hadoop versions should match the corresponding versions on the remote cluster.

Suppose we only want to compute average arrival delays per carrier for flights with a distance over 1000 miles. Variable selection and filtering of observations is implemented in select() and filter() (as in the dplyr package).

```
# filter
long_flights <- select(flights, "carrier", "year", "arr_delay", "distance")
long_flights <- filter(long_flights, long_flights$distance >= 1000)
head(long_flights)
```

```
##   carrier year arr_delay distance
## 1      UA 2013        11     1400
## 2      UA 2013        20     1416
## 3      AA 2013        33     1089
## 4      B6 2013       -18     1576
## 5      B6 2013        19     1065
## 6      B6 2013        -2     1028
```

Now we summarize the arrival delays for the subset of long flights by carrier. This is the 'split-apply-combine' approach applied in SparkR.

```
# aggregation: mean delay per carrier
long_flights_delays<- summarize(groupBy(long_flights, long_flights$carrier),
                    avg_delay = mean(long_flights$arr_delay))
head(long_flights_delays)
```

```
##   carrier avg_delay
## 1      UA    3.2622
## 2      AA    0.4958
## 3      EV   15.6876
## 4      B6    9.0364
## 5      DL   -0.2394
## 6      OO   -2.0000
```

Finally, we want to convert the result back into a usual data.frame (loaded in our current R session) in order to further process the summary statistics (output to La-TeX table, plot, etc.). Note that as in the previous aggregation exercises with the ff package, the computed summary statistics (in the form of a table/df) are obviously much smaller than the raw data. However, note that converting a SparkDataFrame back into a native R object generally means all the data stored in the RDDs constituting the SparkDataFrame object is loaded into local RAM. Hence, when working with actual Big Data on a Spark cluster, this type of operation can quickly overflow local RAM.

```
# Convert result back into native R object
delays <- collect(long_flights_delays)
class(delays)
```

```
## [1] "data.frame"
```

```
delays
```

```
##    carrier avg_delay
## 1       UA    3.2622
## 2       AA    0.4958
## 3       EV   15.6876
## 4       B6    9.0364
## 5       DL   -0.2394
## 6       OO   -2.0000
## 7       F9   21.9207
## 8       US    0.5567
## 9       MQ    8.2331
## 10      HA   -6.9152
## 11      AS   -9.9309
## 12      VX    1.7645
## 13      WN    9.0842
## 14      9E    6.6730
```

6.5 Spark with SQL

Instead of interacting with Spark via R, you can do the same via SQL. This can be very convenient at the stage of data exploration and data preparation. Also note that this is a very good example of how knowing some SQL can be very useful when working with Big Data even if you are not interacting with an actual relational database.[7]

To directly interact with Spark via SQL, open a terminal window, switch to the SPARK-HOME directory,

---

[7]Importantly, this also means that we cannot use SQL commands related to configuring such databases, such as .tables etc. Instead we use SQL commands to directly query data from JSON or CSV files.

```
cd SPARK-HOME
```

and enter the following command:

```
$ bin/spark-sql
```

where SPARK-HOME is again the placeholder for the path to your local Spark installation (printed to the console after running SparkR::install.spark()). This will start up Spark and connect to it via Spark's SQL interface. You will notice that the prompt in the terminal changes (similar to when you start sqlite).

Let's run some example queries. The Spark installation comes with several data and script examples. The example datasets are located at SPARK-HOME/examples/src/main/resources. For example, the file employees.json contains the following records in JSON format:

```
{"name":"Michael", "salary":3000}
{"name":"Andy", "salary":4500}
{"name":"Justin", "salary":3500}
{"name":"Berta", "salary":4000}
```

We can query this data directly via SQL commands by referring to the location of the original JSON file.

Select all observations

```
SELECT *
FROM json.`examples/src/main/resources/employees.json`
;
```

```
Michael 3000
Andy    4500
Justin  3500
Berta   4000
Time taken: 0.099 seconds, Fetched 4 row(s)
```

Filter observations

```
SELECT *
```

```
FROM json.`examples/src/main/resources/employees.json`
WHERE salary <4000
;
```

```
Michael 3000
Justin  3500
Time taken: 0.125 seconds, Fetched 2 row(s)
```

Compute the average salary

```
SELECT AVG(salary) AS mean_salary
FROM json.`examples/src/main/resources/employees.json`;
```

```
3750.0
Time taken: 0.142 seconds, Fetched 1 row(s)
```

6.6   Spark with R + SQL

Most conveniently, you can combine the SQL query features of Spark and SQL with running R on Spark. First, initiate the Spark session in RStudio and import the data as a Spark data frame.

```
# to install use
# devtools::install_github("cran/SparkR")
# load packages
library(SparkR)
# start session
sparkR.session()
```

```
## Java ref type org.apache.spark.sql.SparkSession id 1
```

```
# read data
flights <- read.df("data/flights.csv", source = "csv", header="true")
```

Now we can make the Spark data frame accessible for SQL queries by register-ing it as a temporary table/view with createOrReplaceTempView() and then run SQL queries on it from within the R session via the sql()-function. sql() will return the results as a Spark data frame (this means the result is also located on the cluster and hardly affects the master node's memory).

```
# register the data frame as a table
createOrReplaceTempView(flights, "flights" )

# now run SQL queries on it
query <-
"SELECT DISTINCT carrier,
year,
arr_delay,
distance
FROM flights
WHERE 1000 <= distance"

long_flights2 <- sql(query)
head(long_flights2)
```

```
##    carrier year arr_delay distance
## 1      DL 2013       -30     1089
## 2      UA 2013       -11     1605
## 3      DL 2013       -42     1598
## 4      UA 2013        -5     1585
## 5      AA 2013         6     1389
## 6      UA 2013       -23     1620
```

6.7   Wrapping up

- At the core of a vertical scaling strategy are so-called *distributed systems* – several computers connected in a network to jointly process large amounts of data.
- In contrast to standard parallel-computing, the different computing nodes in a distributed system do not share the same physical memory. Each of the nodes/-computers in the system has its own CPU, hard disk, and RAM. This architecture requires a different computing paradigm to run the same data analytics job across all nodes (in parallel).
- A commonly used paradigm to do this is MapReduce, which is implemented in software called Apache Hadoop.
- The core idea of MapReduce is to split a problem/computing task on a large dataset into several components, each of which focuses on a smaller subset of the dataset. The task components are then distributed across the cluster, so that

each component is handled by one computer in the network. Finally, each node returns its result to the master node (the computer coordinating all activities in the cluster), where the partial results are combined into the overall result.

- A typical example of a MapReduce job is the computation of term frequencies in a large body of text. Here, each node computes the number of occurrences of specific words in a subset of the overall body of text; the individual results are then summed up per unique word.
- Apache Hadoop is a collection of open-source software tools to work with massive amounts of data on a distributed system (a network of computers). Part of Hadoop is the Hadoop MapReduce implementation to run MapReduce jobs on a Hadoop cluster.
- Apache Spark is an analytics engine for large-scale data processing on local machines or clusters. It improves upon several shortcomings of the previous Hadoop/MapReduce framework, in particular with regard to iterative tasks (such as in machine learning).

7

Cloud Computing

In this chapter, we first look at what cloud computing basically is and what platforms provide cloud computing services. We then focus on *scaling up* in the cloud. For the sake of simplicity, we will primarily focus on how to use cloud instances provided by one of the providers, Amazon Web Services (AWS). However, once you are familiar with setting things up on AWS, also using Google Cloud, Azure, etc. will be easy. Most of the core services are provided by all providers, and once you understand the basics, the different dashboards will look quite familiar. In a second step, we look at a prominent approach to *scaling out* by setting up a Spark cluster in the cloud.

7.1 Cloud computing basics and platforms

So far we have focused on the available computing resources on our local machines (desktop/laptop) and how to use them optimally when dealing with large amounts of data and/or computationally demanding tasks. A key aspect of this has been to understand why our local machine is struggling with a computing task when there is a large amount of data to be processed and then identifying potential avenues to use the available resources more efficiently, for example, by using one of the following approaches:

- Computationally intensive tasks (but not pushing RAM to the limit): parallelization, using several CPU cores (nodes) in parallel.
- Memory-intensive tasks (data still fits into RAM): efficient memory allocation.
- Memory-intensive tasks (data does not fit into RAM): efficient use of virtual memory (use parts of mass storage device as virtual memory).
- Storage: efficient storage (avoid redundancies).

In practice, datasets might be too large for our local machine even if we take all of the techniques listed above into account. That is, a parallelized task might still take ages to complete because our local machine has too few cores available, a task involving virtual memory would use up way too much space on our hard disk, etc.

In such situations, we have to think about horizontal and vertical scaling beyond our local machine. That is, we outsource tasks to a bigger machine (or a cluster of machines) to which our local computer is connected (typically, over the internet). While only one or two decades ago most organizations had their own large centrally hosted machines (database servers, cluster computers) for such tasks, today they often rely on third-party solutions *'in the cloud'*. That is, specialized companies provide computing resources (usually, virtual servers) that can be easily accessed via a broadband internet connection and rented on an hourly basis (or even by the minute or second). Given the obvious economies of scale in this line of business, a few large players have emerged who effectively dominate most of the global market:

- Amazon Web Services (AWS)[1]
- Microsoft Azure[2]
- Google Cloud Platform (GCP)[3]
- IBM Cloud[4]
- Alibaba Cloud[5]
- Tencent Cloud[6]

In the following subsections and chapters, we will primarily rely on services provided by AWS and GCP. In order to try out the code examples and tutorials, make sure to have an AWS account as well as a Google account (which can then easily be linked to GCP). For the AWS account, go to `https://aws.amazon.com/` and create an account. You will have to enter credit card details for either cloud platform when setting up/linking accounts. Importantly, you will only be charged for the time you use an AWS service. However, even when using some cloud instances, several of AWS's cloud products offer a free tier to test and try out products. The following examples rely whenever possible on free-tier instances; if not, it is explicitly indicated that running the example in the cloud will generate some costs on your account. For the GCP account, have your Google login credentials ready, and visit `https://cloud.google.com/` to register your Google account with GCP. Again, credit card details are needed to set up an account, but many of the services can be used for free to a certain extent (to learn and try out code).

---

[1]https://aws.amazon.com/
[2]https://azure.microsoft.com/en-us/
[3]https://cloud.google.com/
[4]https://www.ibm.com/cloud/
[5]https://www.alibabacloud.com/
[6]https://intl.cloud.tencent.com/

7.2   Transitioning to the cloud

When logged in to AWS and GCP, you will notice the breadth of services offered by these platforms. There are more than 10 main categories of services, with many subcategories and products in each. It is easy to get lost from just browsing through them. Rest assured that for the purpose of data analytics/applied econometrics, many of these services are irrelevant. Our motivation to use the cloud is to extend our computational resources to use our analytics scripts on large datasets, not to develop and deploy web applications or business analytics dashboards. With this perspective, a small selection of services will make the cloud easily accessible for daily analytics workflows.

When we use services from AWS or GCP to *scale up* (vertical scaling) the available resources, the transition from our local implementation of a data analytics task to the cloud implementation is often rather simple. Once we have set up a cloud instance and figured out how to communicate with it, we typically can run the exact same R script locally and in the cloud. This is usually the case for parallelized tasks (simply run the same script on a machine with more cores), in-memory tasks (rent a machine with more RAM but still use `data.table()`, etc.), and highly parallelized tasks to be run on GPUs. The transition from a local implementation to horizontal scaling (*scaling out*) in the cloud will require slightly more preparatory steps. However, in this domain we will directly build on the same (or very similar) software tools that we have used locally in previous chapters. For example, instead of connecting R to a local SQLite database, we set up a MySQL database on AWS RDS and then connect in essentially the same way our local R session with this database in the cloud.

7.3   Scaling up in the cloud: Virtual servers

In the following pages we look at a very common scheme to deal with a lack of local computing resources: flexibly renting a type of virtual server often referred to as "Elastic Cloud Computing (EC2)" instance. Specifically, we will look at how to scale up with AWS EC2 and R/RStudio Server. One of the easiest ways to set up an AWS EC2 instance for R/RStudio Server is to use Louis Aslett's Amazon Machine Image (AMI)[7]. This way you do not need to install R/Rstudio Server yourself. Simply follow these five steps:

---

[7]https://www.louisaslett.com/RStudio_AMI/

- Depending on the region in which you want to create your EC2 instance, click on the corresponding AMI link in https://www.louisaslett.com/RStudio_AMI/. For example, if you want to create the instance in Frankfurt, click on ami-076abd591c4335092[8]. You will be automatically directed to the AWS page where you can select the type of EC2 instance you want to create. By default, the free tier T2.micro instance is selected (I recommend using this type of instance if you simply want to try out the examples below).

- After selecting the instance type, click on "Review and Launch". On the opened page, select "Edit security groups". There should be one entry with ssh selected in the drop-down menu. Click on this drop-down menu and select HTTP (instead of ssh). Click again on "Review and Launch" to confirm the change.

- Then, click "Launch" to initialize the instance. From the pop-up concerning the key pair, select "Proceed without a key pair" from the drop-down menu, and check the box below ("I acknowledge ..."). Click "Launch" to confirm. A page opens. Click on "View" instances to see all of your instances and their status. Wait until "Status check" is "2/2 checks passed" (you might want to refresh the instance overview or browser window).

- Click on the instance ID of your newly launched instance and copy the public IPv4 address, open a new browser window/tab, type in `http://`, paste the IP address, and hit enter (the address in your browser bar will be something like `http://3.66.120.150; http,` not `https!`).

- You should see the login-interface to RStudio on your cloud instance. The username is `rstudio`, and the password is the instance ID of your newly launched instance (it might take a while to load R/Rstudio). Once RStudio is loaded, you are ready to go.

NOTE: the instructions above help you set up your own EC2 instance with R/RStudio to run some example scripts and tryout R on EC2. For more serious/professional (long-term) usage of an EC2 instance, I strongly recommend setting it up manually and improving the security settings accordingly! The above setup will theoretically result in your instance being accessible for anyone in the Web (something you might want to avoid).

7.3.1 Parallelization with an EC2 instance

This short tutorial illustrates how to scale the computation up by running it on an AWS EC2 instance. Thereby, we build on the techniques discussed in the previous chapter. Note that our EC2 instance is a Linux machine. When running R on a

---

[8]https://console.aws.amazon.com/ec2/home?region=eu-central-1#launchAmi=ami-076abd59 1c4335092

Linux machine, there is sometimes an additional step to install R packages (at least for most of the packages): R packages need to be compiled before they can be installed. The command to install packages is exactly the same (`install.packages()`), and normally you only notice a slight difference in the output shown on the R console during installation (and the installation process takes a little longer than what you are used to). In some cases you might also have to install additional dependencies directly in Linux. Apart from that, using R via RStudio Server in the cloud looks/feels very similar if not identical to when using R/RStudio locally.

Preparatory steps

If your EC2 instance with RStudio Server is not running yet, do the following. In the AWS console, navigate to EC2, select your EC2 instance (with RStudio Server installed), and click on "Instance state/Start instance". You will have to wait until you see "2/2 checks passed". Then, open a new browser window, enter the address of your EC2/RStudio Server instance (see above, e.g., `http://3.66.120.150`), and log in to RStudio. First, we need to install the `parallel` (R Core Team, 2021) and `doSNOW` (Microsoft Corporation and Weston, 2022) packages. In addition we will rely on the `stringr` package (Wickham, 2022b).

```
# install packages for parallelization
install.packages("parallel", "doSNOW", "stringr")
```

Once the installations have finished, you can load the packages and verify the number of cores available on your EC2 instance as follows. If you have chosen the free tier T2.micro instance type when setting up your EC2 instance, you will see that you only have one core available. Do not worry. It is reasonable practice to test your parallelization script with a few iterations on a small machine before bringing out the big guns. The specialized packages we use for parallelization here do not mind if you have one or 32 cores; the same code runs on either machine (obviously not very fast with only one core).

```
# load packages
library(parallel)
library(doSNOW)

# verify no. of cores available
n_cores <- detectCores()
n_cores
```

Finally, we have to upload the data that we want to process as part of the parallelization task. To this end, in RStudio Server, navigate to the file explorer in the

lower right-hand corner. The graphical user interfaces of a local RStudio installation and RStudio Server are almost identical. However, you will find in the file
explorer pane an "Upload" button to transfer files from your local machine to the
EC2 instance. In this demonstration, we will work with the previously introduced
`marketing_data.csv` dataset. You can thus click on "Upload" and upload it to the current target directory (the home directory of RStudio Server). As soon as the file
is uploaded, you can work with it as usual (as on the local RStudio installation). To
keep things as in the local examples, use the file explorer to create a new `data` folder,
and move `marketing_data.csv` in this new folder. The screenshot in Figure 7.1 shows
a screenshot of the corresponding section.

FIGURE 7.1: File explorer and Upload button on RStudio Server.

In order to test if all is set up properly to run in parallel on our EC2 instance, open a
new R script in RStudio Server and copy/paste the preparatory steps and the simple
parallelization example from Section 4.5 into the R script.

```
# PREPARATION ----------------------------

# packages
library(stringr)

# import data
marketing <- read.csv("data/marketing_data.csv")
# clean/prepare data
marketing$Income <- as.numeric(gsub("[[:punct:]]", "", marketing$Income))
marketing$days_customer <- as.Date(Sys.Date())-
  as.Date(marketing$Dt_Customer, "%m/%d/%y")
marketing$Dt_Customer <- NULL

# all sets of independent vars
indep <- names(marketing)[ c(2:19, 27,28)]
combinations_list <- lapply(1:length(indep),
                            function(x) combn(indep, x, simplify = FALSE))
combinations_list <- unlist(combinations_list, recursive = FALSE)
```

```
models <- lapply(combinations_list,
                function(x) paste("Response ~", paste(x, collapse="+")))
```

Test parallelized code

Now, we can start testing the code on EC2 without registering the one core for cluster processing. This way, %dopar% will automatically resort to running the code sequentially. Make sure to set N to 10 (or another small number) for this test.

```
# set cores for parallel processing
# ctemp <- makeCluster(ncores)
# registerDoSNOW(ctemp)

# prepare loop
N <- 10 # just for illustration, the actual code is N <- length(models)
# run loop in parallel
pseudo_Rsq <-
   foreach ( i = 1:N, .combine = c) %dopar% {
     # fit the logit model via maximum likelihood
     fit <- glm(models[[i]], data=marketing, family = binomial())
     # compute the proportion of deviance explained
     #by the independent vars (~R^2)
     return(1-(fit$deviance/fit$null.deviance))
}
```

Once the test has run through successfully, we are ready to scale up and run the
actual workload in parallel in the cloud.

Scale up and run in parallel

First, switch back to the AWS EC2 console and stop the instance by selecting the
tick-mark in the corresponding row, and click on "Instance state/stop instance".
Once the Instance state is "Stopped", click on "Actions/Instance settings/change
instance type". You will be presented with a drop-down menu from which you can
select the new instance type and confirm. The example below is based on selecting the t2.2xlarge (with 8 vCPU cores and 32MB of RAM). Now you can start the
instance again, log in to RStudio Server (as above), and run the script again – but
this time with the following lines not commented out (in order to make use of all
eight cores):

```
# set cores for parallel processing
ctemp <- makeCluster(ncores)
registerDoSNOW(ctemp)
```

In order to monitor the usage of computing resources on your instance, switch to the Terminal tab, type in `htop`, and hit enter. This will open the interactive process viewer called htop[9]. Figure 7.2 shows the output of htop for the preparatory phase of the parallel task implemented above. The output confirms the available resources provided by a `t2.2xlarge` EC2 instance (with 8 vCPU cores and 32MB of RAM). When using the default free tier T2.micro instance, you will notice in the htop output that only one core is available.

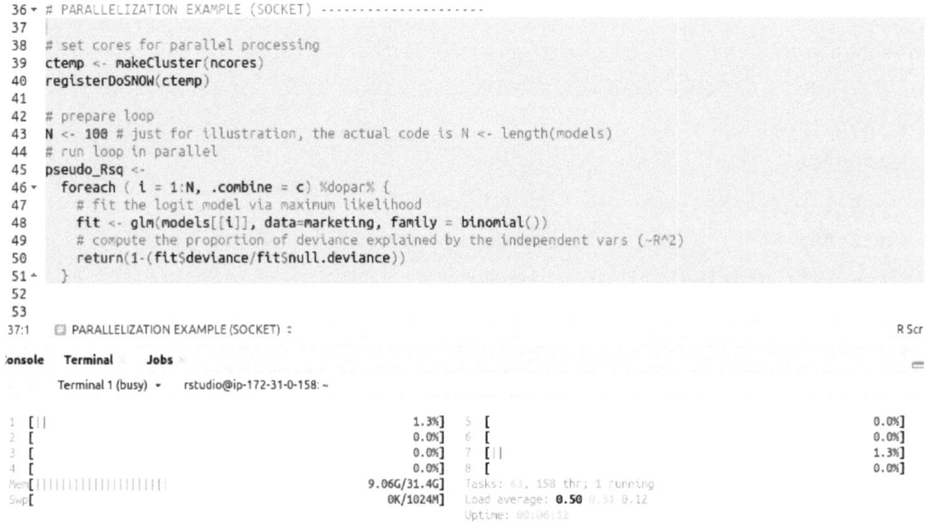

FIGURE 7.2: Monitor resources and processes with htop.

7.4   Scaling up with GPUs

As discussed in Chapter 4, GPUs can help speed up highly parallelizable tasks such as matrix multiplications. While using a local GPU /graphics card for statistical analysis has become easier due to more easily accessible software layers around the GPUs , it still needs solid knowledge regarding the installation of specific GPU drivers and changing of basic system settings. Many users specializing in the data analytics side rather than the computer science/hardware side of Big Data Analytics might not be comfortable with making such installations/changes on their

---

[9]https://htop.dev/

desktop computers or might not have the right type of GPU/graphics card in their device for such changes. In addition, for many users it might not make sense to have a powerful GPU in their local machine, if they only occasionally use it for certain machine learning or parallel computing tasks. In recent years, many cloud computing platforms have started providing virtual machines with access to GPUs , in many cases with additional layers of software and/or pre-installed drivers, allowing users to directly run their code on GPUs in the cloud. Below, we briefly look at two of the most easy-to-use options to run code on GPUs in the cloud: using Google Colab notebooks with GPUs and setting up RStudio on virtual machines in a special EC2 tier with GPU access on AWS.

7.4.1 GPUs on Google Colab

Google Colab provides a very easy way to run R code on GPUs from Google Cloud. All you need is a Google account. Open a new browser window, go to https://colab.to/r, and log in with your Google account if prompted to do so. Colab will open a Jupyter notebook[10] with an R runtime. Click on "Runtime/Change runtime type", and in the drop-down menu under 'Hardware accelerator', select the option 'GPU'.

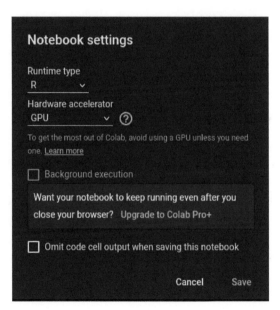

FIGURE 7.3: Colab notebook with R runtime and GPUs.

Then, you can install the packages for which you wish to use GPU acceleration (e.g., `gpuR`, `keras`, and `tensorflow`), and the code relying on GPU processing will be run on GPUs (or even TPUs[11]). At the following link you can find a Colab

---

[10]https://en.wikipedia.org/wiki/Project_Jupyter
[11]https://en.wikipedia.org/wiki/Tensor_Processing_Unit

notebook set up for running a simple image classification tutorial[12] with keras on TPUs: bit.ly/bda_colab[13].

7.4.2   RStudio and EC2 with GPUs on AWS

To start a ready-made EC2 instance with GPUs and RStudio installed, open a browser window and navigate to this service provided by Inmatura on the AWS Marketplace: https://aws.amazon.com/marketplace/pp/prodview-p4gqghzifh mmo. Click on "Continue to Subscribe".

FIGURE 7.4: JupyterHub AMI provided by Inmatura on the AWS Marketplace to run RStudio Server with GPUs on AWS EC2.

After the subscription request is processed, click on "Continue to Configuration" and "Continue to Launch". To make use of a GPU, select, for example, `g2.2xlarge` type under "EC2 Instance Type". If necessary, create a new key pair under Key Pair Settings; otherwise keep all the default settings as they are. Then, at the bottom, click on *Launch*. This will launch a new EC2 instance with a GPU and with RStudio server (as part of JupyterHub) installed.[14]

Once you have successfully launched your EC2 instance, JupyterHub is programmed to automatically initiate on port 80. You can access it using the following link: http://, where the `<instance-ip>` is the public IP address of the newly launched instance (you will find this on the EC2 dashboard). The default username is set as 'jupyterhub-admin', and the default password is identical to your EC2 instance ID. If you need to verify this, you can find it in your EC2 dashboard. For example, it could appear similar to 'i-0b3445939c7492'.[15]

---

[12]https://tensorflow.rstudio.com/tutorials/beginners/basic-ml/tutorial_basic_classification/
[13]https://bit.ly/bda_colab
[14]Note that due to high demand for GPUs on AWS, you might not be able to launch the instance of the preferred type in the preferred region. You will see a corresponding error message after clicking on launch. It might well be the case that simply navigating back to the Configuration page and changing the region of the instance resolves this issue (as not all instances of the preferred type might be in use in other regions).
[15]It is strongly recommended to change the password afterward. In order to do that, click on "Tools" and select "shell". Then type "password" into the shell/terminal, and enter the current password (the instance_id); then enter the new password, hit enter, and enter the new password again to confirm. For further information or help, consult the comprehensive documentation available at: https://aws.inmatura.com/ami/jupyterhub/.

FIGURE 7.5: Launch JupyterHub with RStudio Server and GPUs on AWS EC2.

7.5 Scaling out: MapReduce in the cloud

Many cloud computing providers offer specialized services for MapReduce tasks in the cloud. Here we look at a comparatively easy-to-use solution provided by AWS, called Elastic MapReduce (AWS EMR). It allows you to set up a Hadoop cluster in the cloud within minutes and requires essentially no additional configuration if the cluster is being used for the kind of data analytics tasks discussed in this book.

Setting up a default AWS EMR cluster via the AWS console is straightforward. Simply go to `https://console.aws.amazon.com/elasticmapreduce/`, click on "Create cluster", and adjust the default selection of settings if necessary. Alternatively, we can set up an EMR cluster via the AWS command-line interface (CLI). In the following tutorials, we will work with AWS EMR via R/Rstudio (specifically, via the package `sparklyr`). By default, RStudio is not part of the EMR cluster set-up. However, AWS EMR offers a very flexible way to install/configure additional software on virtual EMR clusters via so-called "bootstrap" scripts. These scripts can be shared on AWS S3 and used by others, which is what we do in the following cluster set-up via the AWS command-line interface (CLI).[16]

---

[16] Specifically, we will use the bootstrap script provided by the AWS Big Data Blog, which is stored here: s3://aws-bigdata-blog/artifacts/aws-blog-emr-rstudio-sparklyr/rstudio_sparklyr_emr6.sh

In order to run the cluster set up via AWS CLI, shown below, you need an SSH key to later connect to the EMR cluster. If you do not have such an SSH key for AWS yet, follow these instructions to generate one: https://docs.aws.amazon.com/clou dhsm/classic/userguide/generate_ssh_key.html. In the example below, the key generated in this way is stored in a file called `sparklyr.pem`.[17]

The following command (`aws emr create-cluster`) initializes our EMR cluster with a specific set of options (all of these options can also be modified via the AWS console in the browser). `--applications Name=Hadoop Name=Spark Name=Hive Name=Pig Name=Tez Name=Ganglia` specifies which type of basic applications (that are essential to running different types of MapReduce tasks) should be installed on the cluster. Unless you really know what you are doing, do not change these settings. `--name "EMR 6.1 RStudio + sparklyr` simply specifies what the newly initialized cluster should be called (this name will then appear on your list of clusters in the AWS console). More relevant for what follows is the line specifying what type of virtual servers (EC2 instances) should be used as part of the cluster: `--instance-groups InstanceGroupType=MASTER,InstanceCount=1,InstanceType=m3.2xlarge` specifies that the one master node (the machine distributing tasks and coordinating the MapReduce procedure) is an instance of type `m3.2xlarge`; `InstanceGroupType=CORE,InstanceCount=2,InstanceType=m3.2xlarge` specifies that there are two slave nodes in this cluster, also of type `m1.medium`.[18] `--bootstrap-action Path=s3://aws-bigdata-blog/artifacts/aws-blog-emr-rstudio-sparklyr/rstudio _sparklyr_emr6.sh,Name="Install RStudio"` tells the set-up application to run the corresponding bootstrap script on the cluster in order to install the additional software (here RStudio).

Finally, there are two important aspects to note: First, in order to initialize the cluster in this way, you need to have an SSH key pair (for your EC2 instances) set up, which you then instruct the cluster to use with `KeyName=`. That is, `KeyName="sparklyr"` means that the user already has created an SSH key pair called `sparklyr` and that this is the key pair that will be used with the cluster nodes for SSH connections. Second, the `--region` argument defines in which AWS region the cluster should be created. Importantly, in this particular case, the bootstrap script used to install RStudio on the cluster is stored in the `us-east-1` region; hence we also need to set up the cluster in this region: `--region us-east-1` (otherwise the set-up will fail as the set-up application will not find the bootstrap script and will terminate with an error!).

---

[17]If you simply copy and paste the CLI command below to set up an EMR cluster, make sure to name your key file `sparklyr.pem`. Otherwise, make sure to change the part in the command referring to the key file accordingly.

[18]Working with one master node of type `m3.2xlarge` and two slave nodes of the same type only makes sense for test purposes. For an actual analysis task with many gigabytes or terabytes of data, you might want to choose larger instances.

```
aws emr create-cluster \
--release-label emr-6.1.0 \
--applications Name=Hadoop Name=Spark Name=Hive Name=Pig \
Name=Tez Name=Ganglia \
--name "EMR 6.1 RStudio + sparklyr"  \
--service-role EMR_DefaultRole \
--instance-groups InstanceGroupType=MASTER,InstanceCount=1,\
InstanceType=m3.2xlarge,InstanceGroupType=CORE,\
InstanceCount=2,InstanceType=m3.2xlarge \
--bootstrap-action \
Path='s3://aws-bigdata-blog/artifacts/
aws-blog-emr-rstudio-sparklyr/rstudio_sparklyr_emr6.sh',\
Name="Install RStudio" --ec2-attributes InstanceProfile=EMR_EC2_DefaultRole,\
KeyName="sparklyr"
--configurations '[{"Classification":"spark",
"Properties":{"maximizeResourceAllocation":"true"}}]' \
--region us-east-1
```

Setting up this cluster with all the additional software and configurations from the bootstrap script will take around 40 minutes. You can always follow the progress in the AWS console. Once the cluster is ready, you will see something like this:

FIGURE 7.6: AWS EMR console indicating the successful set up of the EMR cluster.

In order to access RStudio on the EMR cluster's master node via a secure SSH connection, follow these steps:

- First, follow the prerequisites to connect to EMR via SSH: https://docs.aws.ama zon.com/emr/latest/ManagementGuide/emr-connect-ssh-prereqs.html.

- Then initialize the SSH tunnel to the EMR cluster as instructed here: https://do cs.aws.amazon.com/emr/latest/ManagementGuide/emr-ssh-tunnel.html.

- Protect your key-file (sparklyr.pem) by navigating to the location of the key-file on your computer in the terminal and run chmod 600 sparklyr.pem before connecting. Also make sure your IP address is still the one you have entered in the previous step (you can check your current IP address by visiting https://whatismyipaddr ess.com/).

- In a browser tab, navigate to the AWS EMR console, click on the newly created cluster, and copy the "Master public DNS". In the terminal, connect to the EMR cluster via SSH by running `ssh -i sparklyr.pem -ND 8157 hadoop@master-node-dns` (if you have protected the key-file as superuser, i.e., `sudo chmod`, you will need to use `sudo ssh` here; make sure to replace `master-node-dns` with the actual DNS copied from the AWS EMR console). The terminal will be busy, but you won't see any output (if all goes well).

- In your Firefox browser, install the FoxyProxy add-on[19]. Follow these instructions to set up the proxy via FoxyProxy: https://docs.aws.amazon.com/emr/latest/M anagementGuide/emr-connect-master-node-proxy.html.

- Select the newly created Socks5 proxy in FoxyProxy.

- Go to http://localhost:8787/ and enter with username `hadoop` and password `hadoop`.

Now you can run `sparklyr` on the AWS EMR cluster. After finishing working with the cluster, make sure to terminate it via the EMR console. This will shut down all EC2 instances that are part of the cluster (and hence AWS will stop charging you for this). Once you have connected and logged into RStudio on the EMR cluster's master node, you can connect the Rstudio session to the Spark cluster as follows:

```
# load packages
library(sparklyr)
# connect rstudio session to cluster
sc <- spark_connect(master = "yarn")
```

After using the EMR Spark cluster, make sure to terminate the cluster in the AWS EMR console to avoid additional charges. This automatically terminates all the EC2 machines linked to the cluster.

7.6   Wrapping up

- Cloud computing refers to the on-demand availability of computing resources. While many of today's cloud computing services go beyond the scope of the common data analytics tasks discussed in this book, a handful of specific services

---

[19]https://addons.mozilla.org/en-US/firefox/addon/foxyproxy-standard/

can be very efficient in providing you with the right solution if local computing resources are not sufficient, as summarized in the following bullet points.

- *EC2 (elastic cloud computing)*: scale your analysis up with a virtual server/virtual machine in the cloud. For example, rent an EC2 instance for a couple of minutes in order to run a massively parallel task on 36 cores.
- *GPUs in the cloud*: Google Colab offers an easy-to-use interface to run your machine-learning code on GPUs, for example, in the context of training neural nets.
- *AWS RDS* offers a straightforward way to set up an SQL database in the cloud without any need for database server installation and maintenance.
- *AWS EMR* allows you to flexibly set up and run your Spark/sparkly or Hadoop code on a cluster of EC2 machines in the cloud.

Part III

Components of Big Data Analytics

Introduction

"Men at some time are masters of their fates. The fault, dear Brutus, is not in our stars, but in ourselves, that we are underlings."(Shakespeare, 2020)

Working with the complexity and size of Big Data is rarely as overwhelming and intimidating as when collecting and preparing the data for analysis. Even when working with small datasets, collecting and preparing data for analysis can easily consume more than two-thirds of an analytics project's time. Larger datasets make the time-consuming tasks preceding the actual statistical analysis even more difficult, as the common tools and workflows for dealing with large datasets at the beginning of a project become overwhelmed and rendered useless.

In this part we aim to tackle these challenges by combining the conceptual basics of part I with the platform basics in part II and look at each practical step involved in analyzing Big Data in the context of applied econometrics/business analytics. Thereby, the focus is primarily on how to handle Big Data in order to gather, prepare, combine, filter, visualize, and summarize it for analytics purposes. That is, we cover all core practical tasks necessary before running more sophisticated econometric analyses or machine learning algorithms on the data. Each of these core tasks is summarized in a separate chapter, whereby the sequence of chapters follows a typical workflow/process in applied data science, usually referred to as *data pipeline*.

The data pipeline concept is useful to illustrate and organize the process from gathering/extracting the raw data from various data sources to the final analytics output. The concept of data pipelines thus helps us to structure all steps involved in a meaningful way. Figure 7.7 illustrates the data pipeline idea, which is consistent with the order and content of how the chapters in this book cover all the tasks involved in the first steps of a Big Data Analytics project. Initial, intermediate and final data entities are displayed as nodes in the flow diagram, and tasks/activities between these entities are displayed as arrows.

Data (science) pipeline

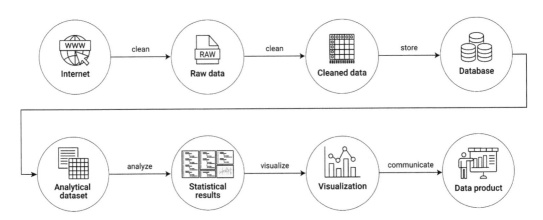

FIGURE 7.7: Illustration of a generic data pipeline in a data analytics project.

In many domains of modern data science, data pipelines also involve loops, indicating that certain parts of the overall process are iterative and repeated several times to further optimize and maintain the final data product. Such a perspective is typical in a context where the final data product is a dashboard or a (component of a) web application that is at some point deployed and runs in production. For the perspective of this book, we intentionally follow an acyclical concept of the data pipeline, representative of data projects in business analytics and applied economic research, in which raw data needs to be collected, processed, and analyzed with the aim of delivering statistical insights summarized in a report or presentation. In that sense, each chapter in this part builds on the previous one, and I recommend going through these chapters sequentially.

Note, though, that in your work it might make sense to slightly re-arrange some parts of the workflow suggested by the generic data pipeline illustrated here. In particular, you might want to first systematically collect and store all of the needed data and then only selectively load and transform/clean parts of this data for analysis (an approach usually referred to as *Extract-Load-Transform (ELT)*, or, in contrast, you might want to directly clean all the gathered data and store all of the cleaned/prepared data in a cleanly structured database before turning to any analytics steps (as the pipeline illustration above would suggest, and which is typically referred to as *Extract-Transform-Load (ETL)*).

Aside: ETL vs. ELT
In traditional business analytics, a company's raw/unstructured data was typically handled and prepared by IT specialists and then provided in a final clean/transformed analytic dataset to the data analysts. Such a process can

generally be described as *Extract-Transform-Load (ETL)*. In ETL the data analyst is only presented with the final well-structured database containing all variables and observations (e.g., in a traditional SQL database). The data pipeline illustration above basically follows the ETL idea in the sense that raw data is gathered/extracted, then cleaned, and then stored in a structured format before it is queried and analyzed. With large amounts of data from various sources with frequent updating (as is typical for Big Data), this process is rather slow and might be inefficient in the sense that parts of the raw data might in the end not play an important role in the data analytics project.

Nowadays, with data analysts and data scientists being increasingly familiar with data technologies and how to handle raw/unstructured data, a more flexible process called *Extract-Load-Transform (ELT)* is followed. In ELT, raw data is extracted and then ingested into/stored in more flexible frameworks than traditional SQL databases that allow for more diverse data structures, such as data warehouses, or do not require the data to be structured, such as data lakes. The transformation of the data for analytics purposes is then done on demand and practically often involves the same person later analyzing the data.

Importantly, for the perspective on data analytics in this book, either process involves the components illustrated in the pipeline above in one way or another. However, the order of the individual steps varies between ETL and ELT.

8

Data Collection and Data Storage

The first steps of a data analytics project typically deal with the question of how to collect, organize, and store the raw data for further processing. In this chapter, we cover several approaches to how to practically implement these steps in the context of observational data, as they commonly occur in applied econometrics and business analytics. Thereby, the focus lies on several important aspects of how to implement these steps locally and then introduces several useful cloud tools to store and query large amounts of data for analytics purposes.

8.1 Gathering and compilation of raw data

The NYC Taxi & Limousine Commission (TLC) provides detailed data on all trip records, including pick-up and drop-off times/locations. When combining all available trip records from 2009 to 2018, we get a rather large dataset of over 200GB. The code examples below illustrate how to collect and compile the entire dataset. In order to avoid long computing times, the code examples shown below are based on a small sub-set of the actual raw data (however, all examples involving virtual memory are in theory scalable to the extent of the entire raw dataset).

The raw data consists of several monthly Parquet files and can be downloaded via the TLC's website[1]. The following short R script automates the downloading of all available trip-record files. *NOTE:* Downloading all files can take several hours and will occupy over 200GB!

```
# Fetch all TLC trip records Data source:
# https://www1.nyc.gov/site/tlc/about/tlc-trip-record-data.page
# Input: Monthly Parquet files from urls


# SET UP -----------------


# packages
```

---

[1]https://www1.nyc.gov/site/tlc/about/tlc-trip-record-data.page

```r
library(R.utils)  # to create directories from within R


# fix vars
BASE_URL <- "https://d37ci6vzurychx.cloudfront.net/trip-data/"
FILE <- "yellow_tripdata_2018-01.parquet"
URL <- paste0(BASE_URL, FILE)
OUTPUT_PATH <- "data/tlc_trips/"
START_DATE <- as.Date("2009-01-01")
END_DATE <- as.Date("2018-06-01")



# BUILD URLS -----------

# parse base url
base_url <- gsub("2018-01.parquet", "", URL)
# build urls
dates <- seq(from = START_DATE, to = END_DATE, by = "month")
year_months <- gsub("-01$", "", as.character(dates))
data_urls <- paste0(base_url, year_months, ".parquet")
data_paths <- paste0(OUTPUT_PATH, year_months, ".parquet")

# FETCH AND STACK CSVS ----------------

mkdirs(OUTPUT_PATH)
# download all csvs in the data range
for (i in 1:length(data_urls)) {

  # download to disk
  download.file(data_urls[i], data_paths[i])
}
```

8.2   Stack/combine raw source files

In a next step, we parse and combine the downloaded data. Depending on how you want to further work with the gathered data, one or another storage format might be more convenient. For the sake of illustration (and the following examples building on the downloaded data), we store the downloaded data in one CSV file. To this end, we make use of the arrow package (Richardson et al., 2022), an R interface to the Apache Arrow C++ library (a platform to work with large-scale columnar data). The aim of the exercise is to combine the downloaded Parquet files into one compressed CSV file, which will be more easily accessible for some of the libraries used in further examples.

We start by installing the `arrow` package in the following way.

```r
# install arrow
Sys.setenv(LIBARROW_MINIMAL = "false") # to enable working with compressed files
install.packages("arrow") # might take a while
```

The setting `LIBARROW_MINIMAL= "false"` ensures that the installation of arrow is not restricted to the very basic functionality of the package. Specifically, for our context it will be important that the `arrow` installation allows for the reading of compressed files.

```r
# SET UP -------------------------

# load packages
library(arrow)
library(data.table)
library(purrr)

# fix vars
INPUT_PATH <- "data/tlc_trips/"
OUTPUT_FILE <- "data/tlc_trips.parquet"
OUTPUT_FILE_CSV <- "data/tlc_trips.csv"

# list of paths to downloaded Parquet files
all_files <- list.files(INPUT_PATH, full.names = TRUE)

# LOAD, COMBINE, STORE ---------------------

# read Parquet files
all_data <- lapply(all_files, read_parquet, as_data_frame = FALSE)

# combine all arrow tables into one
combined_data <- lift_dl(concat_tables)(all_data)

# write combined dataset to csv file
write_csv_arrow(combined_data,
```

```
      file = OUTPUT_FILE_CSV,
      include_header = TRUE)
```

Note that in the code example above we use `purr::lift_dl()` to facilitate the code. The `arrow` function `concat_tables` combines several table objects into one table.

Aside: CSV import and memory allocation, read.csv vs. fread

The time needed for the simple step of importing rather large CSV files can vary substantially in R, depending on the function/package used. The reason is that there are different ways to allocate RAM when reading data from a CSV file. Depending on the amount of data to be read in, one or another approach might be faster. We first investigate the RAM allocation in R with `mem_change()` and `mem_used()`.

```
# SET UP ----------------


# fix variables
DATA_PATH <- "data/flights.csv"
# load packages
library(pryr)
# check how much memory is used by R (overall)
mem_used()

## 1.73 GB


# DATA IMPORT ---------------
# check the change in memory due to each step
# and stop the time needed for the import
system.time(flights <- read.csv(DATA_PATH))

##     user  system elapsed
##    1.825   0.177   2.006


mem_used()

## 1.76 GB
```

```
# DATA PREPARATION --------
flights <- flights[,-1:-3]
# check how much memory is used by R now
mem_used()
```

```
## 1.76 GB
```

The last result is rather interesting. The object flights must have been larger right after importing it than at the end of the script. We have thrown out several variables, after all. Why does R still use that much memory? R does not by default 'clean up' memory unless it is really necessary (meaning no more memory is available). In this case, R still has much more memory available from the operating system; thus there is no need to 'collect the garbage' yet. However, we can force R to collect the garbage on the spot with gc(). This can be helpful to better keep track of the memory needed by an analytics script.

```
gc()
```

```
##              used    (Mb) gc trigger    (Mb)  max used
## Ncells    7028873   375.4   11817067   631.1  11817067
## Vcells 170442328  1300.4  399523409  3048.2 399265766
##              (Mb)
## Ncells     631.1
## Vcells    3046.2
```

Now, let's see how we can improve the performance of this script with regard to memory allocation. Most memory is allocated when importing the file. Obviously, any improvement of the script must still result in importing all the data. However, there are different ways to read data into RAM. read.csv() reads all lines of a csv file consecutively. In contrast, data.table::fread() first 'maps' the data file into memory and only then actually reads it in line by line. This involves an additional initial step, but the larger the file, the less relevant is this first step in the total time needed to read all the data into memory. By switching on the verbose option, we can actually see what fread is doing.

```
# load packages
library(data.table)
```

```
# DATA IMPORT ---------------
system.time(flights <- fread(DATA_PATH, verbose = TRUE))
```

```
##      user  system elapsed
##      0.392   0.004   0.072
```

The output displayed on the console shows what is involved in steps [1] to [12] of the parsing/import procedure. Note in particular the following line under step [7] in the procedure:

```
Estimated number of rows: 30960501 / 92.03 = 336403
  Initial alloc = 370043 rows (336403 + 9%) using
  bytes/max(mean-2*sd,min) clamped between [1.1*estn, 2.0*estn]
```

This is the result of the above-mentioned preparatory step in the form of sampling. The `fread` CSV parser first estimates how large the dataset likely is and then creates an additional allocation (in this case of 370043 rows). Only after this are the rows actually imported into RAM. The summary of the time allocated for the different steps shown at the bottom of the output nicely illustrates that the preparatory steps of memory mapping and allocation are rather fast compared with the time needed to actually read the data into RAM. Given the size of the dataset, `fread`'s approach to memory allocation results in a much faster import of the dataset than `read.csv`'s approach.

8.3   Efficient local data storage

In this section, we are concerned with a) how we can store large datasets permanently on a mass storage device in an efficient way (here, efficient can be understood as 'not taking up too much space') and b) how we can load (parts of) this dataset in an efficient way (here, efficient~fast) for analysis.

We look at this problem in two situations:

- The data needs to be stored locally (e.g., on the hard disk of our laptop).
- The data can be stored on a server 'in the cloud'.

Various tools have been developed over the last few years to improve the efficiency of storing and accessing large amounts of data, many of which go beyond the scope implied by this book's perspective on applied data analytics. Here, we focus on the basic concept of *SQL/Relational Database Management Systems (RDBMSs)*, as well as a few alternatives that can be summarized under the term *NoSQL ('non-SQL', sometimes 'Not only SQL')* database systems. Conveniently (and contrary to what the latter name would suggest), most of these tools can be worked with by using basic SQL queries to load/query data.

The relational database system follows the relational data model, in which the data is organized in several tables that are connected via some unique data record identifiers (keys). Such systems, for example, SQLite introduced in Chapter 3, have been used for a long time in all kinds of business and analytics contexts. They are well-tried and stable and have a large and diverse user base. There are many technicalities involved in how they work under the hood, but for our purposes three characteristics are most relevant:

1. All common RDBMSs, like SQLite and MySQL, are *row-based* databases. That is, data is thought of as observations/data records stored in rows of a table. One record consists of one row.
2. They are typically made for storing clean data in a *clearly defined set of tables*, with clearly defined properties. The organizing of data in various tables has (at least for our perspective here) the aim of avoiding redundancies and thereby using the available storage space more efficiently.
3. Rows are *indexed* according to the unique identifiers of tables (or one or several other variables in a table). This allows for fast querying of specific records and efficient merging/joining of tables.

While these particular features work very well also with large amounts of data, particularly for exploration and data preparation (joining tables), in the age of Big Data they might be more relevant for operational databases (in the back-end of web applications, or simply the operational database of a business) than for the specific purpose of data analytics.

On the one hand, the data basis of an analytics project might be simpler in terms of the number of tables involved. On the other hand, Big Data, as we have seen, might come in less structured and/or more complex forms than traditional table-like/row-based data. *NoSQL* databases have been developed for the purposes of storing more complex/less structured data, which might not necessarily be described as a set of tables connected via keys, and for the purpose of fast analysis of large amounts of data. Again, three main characteristics of these types of databases are of particular relevance here:

1. Typically, *NoSQL* databases are not row-based, but follow a *column-based*, document-based, key-value-based, or graph-based data model. In what follows, the column-based model is most relevant.
2. *NoSQL* databases are designed for horizontal scaling. That is, scaling such a database out over many nodes of a computing cluster is usually straightforward.
3. They are optimized to give quick answers based on summarizing large amounts of data, such as frequency counts and averages (sometimes by using approximations rather than exact computations.)

Figure 8.1 illustrates the basic concept of row-based vs. column-based data storage.

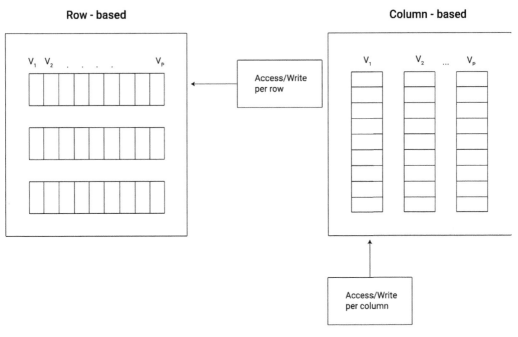

FIGURE 8.1: Schematic illustration of columnar vs. row-based data storage.

Aside: Row-based vs. column-based databases

Conceptually, in a *row-based database* individual values (cells) are contained in rows, which means changing one value requires updating a row. Row-based databases (e.g., SQLite) are thus designed for efficient data reading and writing when users often access many columns but rather few observations. For example, for an operational database in the back-end of a web application such as an online shop, a row-based approach makes sense because hundreds or thousands of users (customers in that case) constantly add or query small amounts of data. In contrast, changing a value in *column-based* databases means changing a column. Accessing all values in a particular column is much faster in comparison to row-based databases.

This means that column-based databases are useful when users tend to query rather few columns but massive numbers of observations, which is typically rather the case in an analytics context. Some well-known data warehouse and data lake systems are therefore based on this principle (e.g., Google BigQuery). However, if analytics tasks involve a lot of (out-of-memory) table joins, column-based solutions are likely to be slower than row-based solutions.

In the following, we have a close look at using both column-based and row-based tools. Thereby we will particularly highlight the practical differences between using column-based and row-based data storage solutions.

8.3.1 RDBMS basics

RDBMSs have two key features that tackle the two efficiency concerns mentioned above:

- The *relational data model*: The overall dataset is split by columns (covariates) into tables in order to reduce the storage of redundant variable-value repetitions. The resulting database tables are then linked via key-variables (unique identifiers). Thus (simply put), each type of entity on which observations exist resides in its own database table. Within this table, each observation has its unique ID. Keeping the data in such a structure is very efficient in terms of storage space used.

- *Indexing*: The key-columns of the database tables are indexed, meaning (in simple terms) ordered on disk. Indexing a table takes time, but it has to be performed only once (unless the content of the table changes). The resulting index is then stored on disk as part of the database. These indices substantially reduce the number of disk accesses required to query/find specific observations. Thus, they make the loading of specific parts of the data for analysis much more efficient.

The loading/querying of data from an RDBMS typically involves the selection of specific observations (rows) and covariates (columns) from different tables. Due to the indexing, observations are selected efficiently, and the defined relations between tables (via keys) facilitate the joining of columns to a new table (the queried data).

8.3.2 Efficient data access: Indices and joins in SQLite

So far we have only had a look at the very basics of writing SQL code. Let us now further explore SQLite as an easy-to-use and easy-to-set-up relational database solution. In a second step we then look at how to connect to a local SQLite database from within R. First, we switch to the Terminal tab in RStudio, set up a new database called air.sqlite, and import the csv-file flights.csv (used in previous chapters) as a first table.

```
# switch to data directory
cd data
# create database and run sqlite
sqlite3 air.sqlite
```

```
-- import csvs
.mode csv
.import flights.csv flights
```

We check whether everything worked out well via the `.tables` and `.schema` commands.

```
.tables
.schema flights
```

In `flights`, each row describes a flight (the day it took place, its origin, its destination, etc.). It contains a covariate `carrier` containing the unique ID of the respective airline/carrier carrying out the flight as well as the covariates `origin` and `dest`. The latter two variables contain the unique IATA-codes of the airports from which the flights departed and where they arrived, respectively. In `flights` we thus have observations at the level of individual flights.

Now we extend our database in a meaningful way, following the relational data model idea. First we download two additional CSV files containing data that relate to the flights table:

- `airports.csv`[2]: Describes the locations of US Airports (relates to `origin` and `dest`).
- `carriers.csv`[3]: A listing of carrier codes with full names (relates to the `carrier`-column in `flights`).

In this code example, the two CSVs have already been downloaded to the `materials/data`-folder.

```
-- import airport data
.mode csv
.import airports.csv airports
.import carriers.csv carriers
```

```
-- inspect the result
.tables
.schema airports
.schema carriers
```

Now we can run our first query involving the relation between tables. The aim of the exercise is to query flights data (information on departure delays per flight number

---

[2]https://bda-examples.s3.eu-central-1.amazonaws.com/airports.csv
[3]https://bda-examples.s3.eu-central-1.amazonaws.com/carriers.csv

and date, from the `flights` table) for all `United Air Lines Inc.` flights (information from the `carriers` table) departing from `Newark Intl` airport (information from the `airports` table). In addition, we want the resulting table ordered by flight number. For the sake of the exercise, we only show the first 10 results of this query (LIMIT 10).

```sql
SELECT
year,
month,
day,
dep_delay,
flight
FROM (flights INNER JOIN airports ON flights.origin=airports.iata)
INNER JOIN carriers ON flights.carrier = carriers.Code
WHERE carriers.Description = 'United Air Lines Inc.'
AND airports.airport = 'Newark Intl'
ORDER BY flight
LIMIT 10;
```

flights_join

```
##      year month day dep_delay flight
## 1   2013     1   4         0      1
## 2   2013     1   5        -2      1
## 3   2013     3   6         1      1
## 4   2013     2  13        -2      3
## 5   2013     2  16        -9      3
## 6   2013     2  20         3      3
## 7   2013     2  23        -5      3
## 8   2013     2  26        24      3
## 9   2013     2  27        10      3
## 10  2013     1   5         3     10
```

Note that this query has been executed without indexing any of the tables first. Thus SQLite could not take any 'shortcuts' when matching the ID columns in order to join the tables for the query output. That is, SQLite had to scan all the columns to find the matches. Now we index the respective ID columns and re-run the query.

```sql
CREATE INDEX iata_airports ON airports (iata);
CREATE INDEX origin_flights ON flights (origin);
```

```
CREATE INDEX carrier_flights ON flights (carrier);
CREATE INDEX code_carriers ON carriers (code);
```

Note that SQLite optimizes the efficiency of the query without our explicit instructions. If there are indices it can use to speed up the query, it will do so.

```
SELECT
year,
month,
day,
dep_delay,
flight
FROM (flights INNER JOIN airports ON flights.origin=airports.iata)
INNER JOIN carriers ON flights.carrier = carriers.Code
WHERE carriers.Description = 'United Air Lines Inc.'
AND airports.airport = 'Newark Intl'
ORDER BY flight
LIMIT 10;
```

```
##      year month day dep_delay flight
## 1   2013     1   4         0      1
## 2   2013     1   5        -2      1
## 3   2013     3   6         1      1
## 4   2013     2  13        -2      3
## 5   2013     2  16        -9      3
## 6   2013     2  20         3      3
## 7   2013     2  23        -5      3
## 8   2013     2  26        24      3
## 9   2013     2  27        10      3
## 10  2013     1   5         3     10
```

You can find the final `air.sqlite` database, including all the indices and tables, as `materials/data/air_final.sqlite` in the book's code repository.

8.4   Connecting R to an RDBMS

The R-package `RSQLite` (Müller et al., 2022) embeds SQLite in R. That is, it provides functions that allow us to use SQLite directly from within R. You will see that the combination of SQLite with R is a simple but very practical approach to working with very efficiently (and locally) stored datasets. In the following example, we explore how RSQLite can be used to set up and query the air.sqlite database shown in the example above.

8.4.1 Creating a new database with RSQLite

Similarly to the raw SQLite syntax, we connect to a database that does not exist yet actually creates this (empty database). Note that for all interactions with the database from within R, we need to refer to the connection (here: con_air).

```
# load packages
library(RSQLite)


# initialize the database
con_air <- dbConnect(SQLite(), "data/air.sqlite")

```

8.4.2 Importing data

With RSQLite we can easily add data.frames as SQLite tables to the database.

```
# import data into current R session
flights <- fread("data/flights.csv")
airports <- fread("data/airports.csv")
carriers <- fread("data/carriers.csv")

# add tables to database
dbWriteTable(con_air, "flights", flights)
dbWriteTable(con_air, "airports", airports)
dbWriteTable(con_air, "carriers", carriers)
```

8.4.3 Issuing queries

Now we can query the database from within R. By default, RSQLite returns the query results as data.frames. Queries are simply character strings written in SQLite.

```
# define query
delay_query <-
"SELECT
year,
month,
```

```
day,
dep_delay,
flight
FROM (flights INNER JOIN airports ON flights.origin=airports.iata)
INNER JOIN carriers ON flights.carrier = carriers.Code
WHERE carriers.Description = 'United Air Lines Inc.'
AND airports.airport = 'Newark Intl'
ORDER BY flight
LIMIT 10;
"

# issue query
delays_df <- dbGetQuery(con_air, delay_query)
delays_df

# clean up
dbDisconnect(con_air)
```

When done working with the database, we close the connection to the database with `dbDisconnect(con)`.

8.5   Cloud solutions for (big) data storage

As outlined in the previous section, RDBMSs are a very practical tool for storing the structured data of an analytics project locally in a database. A local SQLite database can easily be set up and accessed via R, allowing one to write the whole data pipeline – from data gathering to filtering, aggregating, and finally analyzing – in R. In contrast to directly working with CSV files, using SQLite has the advantage of organizing the data access much more efficiently in terms of RAM. Only the final result of a query is really loaded fully into R's memory.

If mass storage space is too sparse or if RAM is nevertheless not sufficient, even when organizing data access via SQLite, several cloud solutions come to the rescue. Although you could also rent a traditional web server and host a SQL database there, this is usually not worthwhile for a data analytics project. In the next section we thus look at three important cases of how to store data as part of an analytics project: *RDBMS in the cloud*, a serverless *data warehouse* solution for large datasets called *Google BigQuery*, and a simple storage service to use as a *data lake* called *AWS*

S3. All of these solutions are discussed from a data analytics perspective, and for all of these solutions we will look at how to make use of them from within R.

8.5.1 Easy-to-use RDBMS in the cloud: AWS RDS

Once we have set up the RStudio server on an EC2 instance, we can run the SQLite examples shown above on it. There are no additional steps needed to install SQLite. However, when using RDBMSs in the cloud, we typically have a more sophisticated implementation than SQLite in mind. Particularly, we want to set up an actual RDBMS server running in the cloud to which several clients can connect (e.g., via RStudio Server).

AWS's Relational Database Service (RDS) provides an easy way to set up and run a SQL database in the cloud. The great advantage for users new to RDBMS/SQL is that you do not have to manually set up a server (e.g., an EC2 instance) and install/configure the SQL server. Instead, you can directly set up a fully functioning relational database in the cloud.

As a first step, open the AWS console and search for/select "RDS" in the search bar. Then, click on "Create database" in the lower part of the landing page.

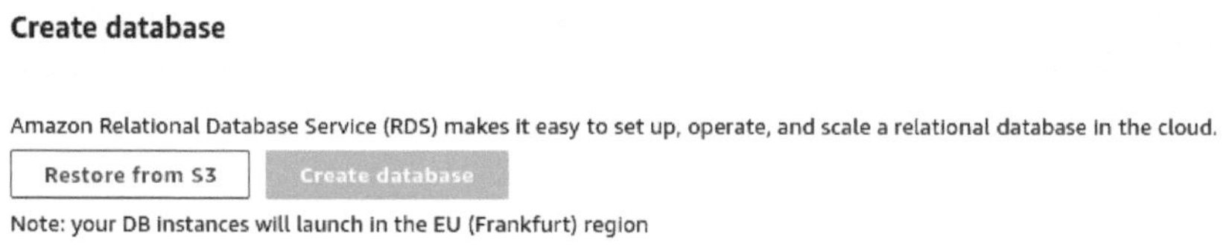

Create database

Amazon Relational Database Service (RDS) makes it easy to set up, operate, and scale a relational database in the cloud.

| Restore from S3 | Create database |

Note: your DB instances will launch in the EU (Frankfurt) region

FIGURE 8.2: Create a managed relational database on AWS RDS.

On the next page, select "Easy create", "MySQL", and the "Free tier" DB instance size. Further down you will have to set the database instance identifier, the user name, and a password.

Once the database instance is ready, you will see it in the databases overview. Click on the DB identifier (the name of your database shown in the list of databases), and click on modify (button in the upper-right corner). In the "Connectivity" panel under "Additional configuration", select *Publicly accessible* (this is necessary to interact with the DB from your local machine), and save the settings. Back on the overview page of your database, under "Connectivity & security", click on the link under the VPC security groups, scroll down and select the "Inbound rules" tab. Edit the inbound rule to allow any IP4 inbound traffic.[4]

---

[4]Note that this is not generally recommended. Only do this to get familiar with the service and to test some code.

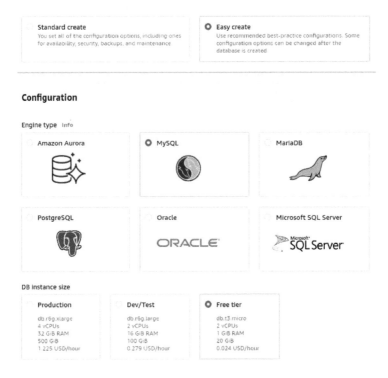

FIGURE 8.3: Easy creation of an RDS MySQL DB.

	Name	Security group r...	IP version	Type	Protocol	Port range	Source
☑	-	sgr-023970b13230...	IPv4	MYSQL/Aurora	TCP	3306	0.0.0.0/0

Inbound rules (1/1)

FIGURE 8.4: Allow all IP4 inbound traffic (set Source to `0.0.0.0/0`).

```
# load packages
library(RMySQL)
library(data.table)

# fix vars
# replace this with the Endpoint shown in the AWS RDS console
RDS_ENDPOINT <- "MY-ENDPOINT"
# replace this with the password you have set when initiating the RDS DB on AWS
PW <- "MY-PW"

# connect to DB
con_rds <- dbConnect(RMySQL::MySQL(),
```

```
                    host=RDS_ENDPOINT,
                    port=3306,
                    username="admin",
                    password=PW)


# create a new database on the MySQL RDS instance
dbSendQuery(con_rds, "CREATE DATABASE air")

# disconnect and re-connect directly to the new DB
dbDisconnect(con_rds)
con_rds <- dbConnect(RMySQL::MySQL(),
                     host=RDS_ENDPOINT,
                     port=3306,
                     username="admin",
                     dbname="air",
                     password=PW)
```

RMySQL and RSQLite both build on the DBI package, which generalizes how we can interact with SQL-type databases via R. This makes it straightforward to apply what we have learned so far by interacting with our local SQLite database to interactions with other databases. As soon as the connection to the new database is established, we can essentially use the same R functions as before to create new tables and import data.

```
# import data into current R session
flights <- fread("data/flights.csv")
airports <- fread("data/airports.csv")
carriers <- fread("data/carriers.csv")

# add tables to database
dbWriteTable(con_rds, "flights", flights)
dbWriteTable(con_rds, "airports", airports)
dbWriteTable(con_rds, "carriers", carriers)
```

Finally, we can query our RDS MySQL database on AWS.

```
# define query
delay_query <-
"SELECT
year,
```

```
month,
day,
dep_delay,
flight
FROM (flights INNER JOIN airports ON flights.origin=airports.iata)
INNER JOIN carriers ON flights.carrier = carriers.Code
WHERE carriers.Description = 'United Air Lines Inc.'
AND airports.airport = 'Newark Intl'
ORDER BY flight
LIMIT 10;
"

# issue query
delays_df <- dbGetQuery(con_rds, delay_query)
delays_df

# clean up
dbDisconnect(con_rds)
```

8.6   Column-based analytics databases

As outlined in the discussion of row-based vs. column-based databases above, many data analytics tasks focus on few columns but many rows, hence making column-based databases the better option for large-scale analytics purposes. Apache Druid[5] (Yang et al., 2014) is one such solution that has particular advantages for the data analytics perspective taken in this book. It can easily be run on a local machine (Linux and Mac/OSX), or on a cluster in the cloud, and it easily allows for connections to external data, for example, data stored on Google Cloud Storage. Moreover, it can be interfaced by RDruid (Metamarkets Group Inc., 2023) to run Druid queries from within R, or, yet again, Druid can be directly queried via SQL.

To get started with Apache Druid, navigate to https://druid.apache.org/. Under downloads[6] you will find a link to download the latest stable release (at the time of writing this book: 25.0.0). On the Apache Druid landing page, you will also find a link Quickstart[7] with all the details regarding the installation and set up.

---

[5]https://druid.apache.org/
[6]https://druid.apache.org/downloads.html
[7]https://druid.apache.org/docs/latest/tutorials/index.html

Importantly, as of the time of writing this book, only Linux and MacOSX are supported (Windows is not supported).

8.6.1 Installation and start up

On Linux, follow these steps to set up Apache Druid on your machine. First, open a terminal and download the Druid binary to the location in which you want to work with Druid. First, we download and unpack the current Apache Druid version via the terminal.

Using Druid in its most basic form is then straightforward. Simply navigate to the unpacked folder and run `./bin/start-micro-quickstart`.

```
# navigate to local copy of druid
cd apache-druid-25.0.0
# start up druid (basic/minimal settings)
./bin/start-micro-quickstart
```

8.6.2 First steps via Druid's GUI

Once all Druid services are running, open a new browser window and navigate to `http://localhost:8888`. This will open Druid's graphical user interface (GUI). The GUI provides easy-to-use interfaces to all basic Druid services, ranging from the loading of data to querying via Druid's SQL. Figure 8.5 highlights the GUI buttons mentioned in the instructions below.

FIGURE 8.5: Apache Druid GUI starting page. White boxes highlight buttons for the Druid services discussed in the main text (from left to right): the query editor (run Druid SQL queries on any of the loaded data sources directly here); the data load service (use this to import data from local files); and the Datasources console (lists all currently available data sources).

8.6.2.1 Load data into Druid

In a first step, we will import the TLC taxi trips dataset from the locally stored CSV file. To do so, click on *Load data/Batch - classic*, then click on *Start new batch spec*, and then select *Local disk* and *Connect*. On the right side of the Druid GUI, a menu will open. In the `Base directory` field, enter the path to the local directory in which you have stored the TLC taxi trips CSV file used in the examples above (`../data/`).[8] In the `File filter` field, enter `tlc_trips.csv`.[9] Finally click on the *Apply* button.

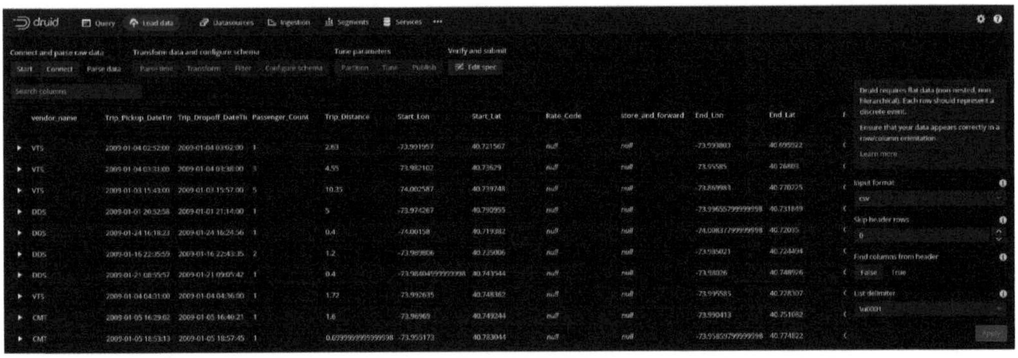

FIGURE 8.6: Apache Druid GUI: CSV parse menu for classic batch data ingestion.

The first few lines of the raw data will appear in the Druid console. In the lower-right corner of the console, click on *Next: Parse data*. Druid will automatically guess the delimiter used in the CSV (following the examples above, this is ,) and present the first few parsed rows.

If all looks good, click on *Next: Parse time* in the lower-right corner of the console. Druid is implemented to work particularly fast on time-series and panel data. To this end, it expects you to define a main time-variable in your dataset, which then can be used to index and partition your overall dataset to speed up queries for specific time frames. Per default, Druid will suggest using the first column that looks like a time format (in the TLC-data, this would be column 2, the pick-up time of a trip, which seems very reasonable for the sake of this example). We move on with a click on *Next: Transform* in the lower right corner. Druid allows you, right at the step of loading data, to add or transform variables/columns. As we do not need to change anything at this point, we continue with *Next: Filter* in the lower-right corner. At this stage you can filter out rows/observations that you are sure should not be included in any of the queries/analyses performed later via Druid.

For this example, we do not filter out any observations and continue via *Next: Configure schema* in the lower-right corner. Druid guesses the schema/data types for each column based on sampling the first few observations in the dataset. Notice,

---

[8]Importantly, recall that with this set up, Druid is currently running from within the `apache-druid-25.0.0`-directory. Hence, unless you have copied your data into this directory, you will have to explicitly point to data files outside of this directory (via `../`).

[9]The Druid service to load data by default allows you to point to various files here via the wildcard character (`*`). As we have stored all the taxi trip example data in one CSV file, we can directly point to this one file.

for example, how Druid considers `vendor_name` to be a `string` and `Trip_distance` to be a `double` (a 64-bit floating point number). In most applications of Druid for the data analytics perspective of this book, the guessed data types will be just fine. We will leave the data types as-is and keep the original column/variable names. You can easily change names of variables/columns by double-clicking on the corresponding column name, which will open a menu on the right-hand side of the console. With this, all the main parameters to load the data are defined. What follows has to do with optimizing Druid's performance.

Once you click on *Next: Partition* in the lower-right corner, you will have to choose the primary partitioning, which is always based on time (again, this has to do with Druid being optimized to work on large time-series and panel datasets). Basically, you need to decide whether the data should be organized into chunks per year, month, week, etc. For this example, we will segment the data according to months. To this end, from the drop-down menu under `Segment granularity`, choose `month`. For the rest of the parameters, we keep the default values. Continue by clicking on *Next: Tune* (we do not change anything here) and then on *Next: Publish*. In the menu that appears, you can choose the name under which the TLC taxi trips data should be listed in the *Datasources* menu on your local Druid installation, once all the data is loaded/processed. Thinking of SQL queries when working with Druid, the `Datasource name` is what you then will use in the FROM statement of a Druid SQL query (in analogy to a table name in the case of RDBMSs like SQLite). We keep the suggested name `tlc_trips`. Thus, you can click on *Edit spec* in the lower-right corner. An editor window will open and display all your load configurations as a JSON file. Only change anything at this step if you really know what you are doing. Finally, click on *Submit* in the lower-right corner. This will trigger the loading of data into Druid. As in the case of the RDBMS covered above, the data ingestion or data loading process primarily involves indexing and writing data to disk. It does not mean importing data to RAM. Since the CSV file used in this example is rather large, this process can take several minutes on a modern laptop computer.

Once the data ingestion is finished, click on the *Datasources* tab in the top menu bar to verify the ingestion. The `tlc_trips` dataset should now appear in the list of data sources in Druid.

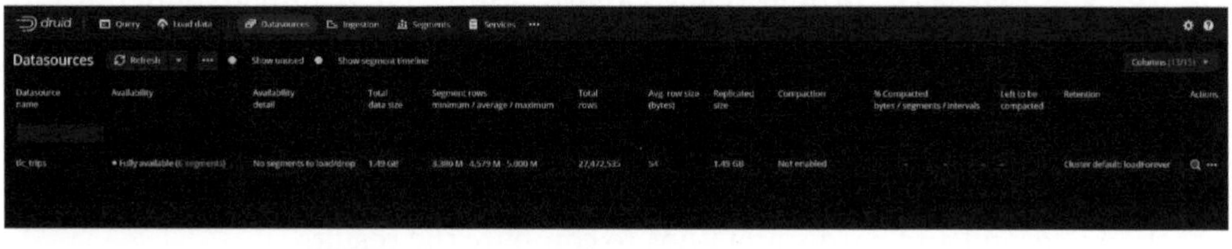

FIGURE 8.7: Apache Druid: Datasources console.

8.6.2.2 Query Druid via the GUI SQL console

Once the data is loaded into Druid, we can directly query it via the SQL console in Druid's GUI. To do this, navigate in Druid to *Query*. To illustrate the strengths of Druid as an analytic database, we run an extensive data aggregation query. Specifically, we count the number of cases (trips) per vendor and split the number of trips per vendor further by payment type.

```
SELECT
vendor_name,
Payment_Type,
COUNT(*) AS Count_trips
FROM tlc_trips
GROUP BY vendor_name, Payment_Type
```

Note that for such simple queries, Druid SQL is essentially identical to the SQL dialects covered in previous chapters and subsections, which makes it rather simple for beginners to start productively engaging with Druid. SQL queries can directly be entered in the query tab; a click on *Run* will send the query to Druid, and the results are shown right below.

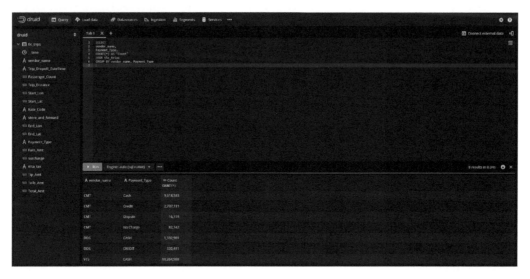

FIGURE 8.8: Apache Druid query console with Druid-SQL example: count the number of cases per vendor and payment type.

Counting the number of taxi trips per vendor name and payment type implies using the entire dataset of over 27 million rows (1.5GB). Nevertheless, Druid needs less than a second and hardly any RAM to compute the results.

8.6.3 Query Druid from R

Apache provides high-level interfaces to Druid for several languages common in data science/data analytics. The RDruid package provides such a Druid connector for R. The package can be installed from GitHub via the devtools package.

```r
# install devtools if necessary
if (!require("devtools")) {
    install.packages("devtools")}


# install RDruid
devtools::install_github("druid-io/RDruid")
```

The RDruid package provides several high-level functions to issue specific Druid queries; however, the syntax might not be straightforward for beginners, and the package has not been further developed for many years.

Thanks to Druid's basic architecture as a web application, however, there is a simple alternative to the RDruid package. Druid accepts queries via HTTP POST calls (with SQL queries embedded in a JSON file sent in the HTTP body). The data is then returned as a compressed JSON string in the HTTP response to the POST request. We can build on this to implement our own simple druid() function to query Druid from R.

```r
# create R function to query Druid (locally)
druid <-
    function(query){
        # dependencies
        require(jsonlite)
        require(httr)
        require(data.table)

        # basic POST body
        base_query <-
        '{
        "context": {
        "sqlOuterLimit": 1001,
        "sqlQueryId": "1"},
        "header": true,
        "query": "",
        "resultFormat": "csv",
        "sqlTypesHeader": false,
```

```
              "typesHeader": false
              }'
          param_list <- fromJSON(base_query)
          # add SQL query
          param_list$query <- query

          # send query; parse result
          resp <- POST("http://localhost:8888/druid/v2/sql",
                       body = param_list,
                       encode = "json")
          parsed <- fread(content(resp, as = "text", encoding = "UTF-8"))
          return(parsed)
      }
```

Now we can send queries to our local Druid installation. Importantly, Druid needs to be started up in order to make this work. In the example below we start up Druid from within R via `system("apache-druid-25.0.0/bin/start-micro-quickstart")` (make sure that the working directory is set correctly before running this). Then, we send the same query as in the Druid GUI example from above.

```
# start Druid
system("apache-druid-25.0.0/bin/start-micro-quickstart",
       intern = FALSE,
       wait = FALSE)
Sys.sleep(30) # wait for Druid to start up

# query tlc data
query <-
'
SELECT
vendor_name,
Payment_Type,
COUNT(*) AS Count_trips
FROM tlc_trips
GROUP BY vendor_name, Payment_Type
'

result <- druid(query)

# inspect result
result
```

```
##      vendor_name Payment_Type Count_trips
## 1:           CMT         Cash     9618583
## 2:           CMT       Credit     2737111
## 3:           CMT      Dispute       16774
## 4:           CMT    No Charge       82142
## 5:           DDS         CASH     1332901
## 6:           DDS       CREDIT      320411
## 7:           VTS         CASH    10264988
## 8:           VTS       Credit     3099625
```

8.7 Data warehouses

Unlike RDBMSs, the main purpose of data warehouses is usually analytics and not the provision of data for everyday operations. Generally, data warehouses contain well-organized and well-structured data, but are not as stringent as RMDBS when it comes to organizing data in relational tables. Typically, they build on a table-based logic, but allow for nesting structures and more flexible storage approaches. They are designed to contain large amounts of data (via horizontal scaling) and are usually column-based. From the perspective of Big Data Analytics taken in this book, there are several suitable and easily accessible data warehouse solutions provided in the cloud. In the following example, we will introduce one such solution called *Google BigQuery*.

8.7.1 Data warehouse for analytics: Google BigQuery example

Google BigQuery is flexible regarding the upload and export of data and can be set up straightforwardly for a data analytics project with hardly any set up costs. The pricing schema is usage-based. Unless you store massive amounts of data on it, you will only be charged for the volume of data processed. Moreover, there is a straightforward R-interface to Google BigQuery called `bigrquery`[10], which allows for the same R/SQL-syntax as R's interfaces to traditional relational databases.

Get started with `bigrquery`

To get started with Google BigQuery and `bigrquery` (Wickham and Bryan, 2022), go to https://cloud.google.com/bigquery. Click on "Try Big Query" (if new to this) or "Go to console" (if used previously). Create a Google Cloud project to use BigQuery with. Note that, as in general for Google Cloud services, you need to have a credit card registered with the project to do this. However, for learning and testing purposes, Google Cloud offers 1TB of free queries per month. All

---

[10]https://bigrquery.r-dbi.org/

the examples shown below combined will not exceed this free tier. Finally, run `install.packages("bigrquery")` in R.

To set up an R session to interface with BigQuery, you need to indicate which Google BigQuery project you want to use for the billing (the BILLING variable in the example below), as well as the Google BigQuery project in which the data is stored that you want to query (the PROJECT variable below). This distinction is very useful because it easily allows you to query data from a large array of publicly available datasets on BigQuery. In the set up example code below, we use this option in order to access an existing and publicly available dataset (provided in the `bigquery-public-data` project) called `google_analytics_sample`. In fact, this dataset provides the raw Google Analytics data used in the Big-P example discussed in Chapter 2.

Finally, all that is left to do is to connect to BigQuery via the already familiar `dbConnect()` function provided in DBI.[11] When first connecting to and querying Big-Query with your Google Cloud account, a browser window will open, and you will be prompted to grant `bigrquery` access to your account/project. To do so, you will have to be logged in to your Google account. See the **Important details** section on https://bigrquery.r-dbi.org/ for details on the authentication.

```
# load packages, credentials
library(bigrquery)
library(data.table)
library(DBI)

# fix vars
# the project ID on BigQuery (billing must be enabled)
BILLING <- "bda-examples"
# the project name on BigQuery
PROJECT <- "bigquery-public-data"
DATASET <- "google_analytics_sample"

# connect to DB on BigQuery
con <- dbConnect(
    bigrquery::bigquery(),
    project = PROJECT,
    dataset = DATASET,
    billing = BILLING
)
```

---

[11]The `bigrquery` package provides a DBI-driver for BigQuery. For more advanced usage of `bigrquery`, the package also provides lower-level functions to directly interact with the BigQuery API.

Get familiar with BigQuery

The basic query syntax is now essentially identical to what we have covered in the RDBMS examples above.[12] In this first query, we count the number of times a Google merchandise shop visit originates from a given web domain on August 1, 2017 (hence the query to table ga_sessions_20170801). Note the way we refer to the specific table (in the FROM statement of the query below): bigquery-public-data is the pointer to the BigQuery project, google_analytics_sample is the name of the data warehouse, and ga_sessions_20170801 is the name of the specific table we want to query data from. Finally, note the argument page_size=15000 as part of the familiar dbGetQuery() function. This ensures that bigrquery does not exceed the limit of volume per second for downloads via the Google BigQuery API (on which bigrquery builds).

```r
# run query
query <-
"
SELECT DISTINCT trafficSource.source AS origin,
COUNT(trafficSource.source) AS no_occ
FROM `bigquery-public-data.google_analytics_sample.ga_sessions_20170801`
GROUP BY trafficSource.source
ORDER BY no_occ DESC;
"
ga <- as.data.table(dbGetQuery(con, query, page_size=15000))
head(ga)
```

Note the output displayed in the console. bigrquery indicates how much data volume was processed as part of the query (which indicates what will be charged to your billing project),

Upload data to BigQuery

Storing your entire raw dataset on BigQuery is straightforward with bigrquery. In the following simple example, we upload the previously gathered and locally stored TLC taxi trips data. To do so, we first create and connect to a new dataset on BigQuery. To keep things simple, we initialize the new dataset in the same project used for the billing.

---

[12]Minor but relevant exceptions are that SQLite, MySQL, and BigQuery do not provide all the same SQL commands. However, for all core operations to query and summarize data, the SQL syntax is essentially identical.

```
# name of the dataset to be created
DATASET <- "tlc"

# connect and initialize a new dataset
con <- dbConnect(
    bigrquery::bigquery(),
    project = BILLING,
    billing = BILLING,
    dataset = DATASET
)
```

In a first step, we create the dataset to which we then can add the table.

```
tlc_ds <- bq_dataset(BILLING, DATASET)
bq_dataset_create(tlc_ds)
```

We then load the TLC dataset into R via `fread()` and upload it as a new table to your project/dataset on BigQuery via `bigrquery`. For the sake of the example, we only upload the first 10,000 rows.

```
# read data from csv
tlc <- fread("data/tlc_trips.csv.gz", nrows = 10000)
# write data to a new table
dbWriteTable(con, name = "tlc_trips", value = tlc)
```

Alternatively, you can easily upload data via the Google BigQuery console in the browser. Go to https://console.cloud.google.com/bigquery, select (or create) the project you want to upload data to, then in the *Explorer* section click on *+ADD DATA*, and select the file you want to upload. You can either upload the data from disk, from Google Cloud Storage, or from a third-party connection. Uploading the data into BigQuery via Google Cloud Storage is particularly useful for large datasets.

Finally, we can test the newly created dataset/table with the following query

```
test_query <-
"
SELECT *
FROM tlc.tlc_trips
LIMIT 10
```

```
"
test <- dbGetQuery(con, test_query)
```

Tutorial: Retrieve and prepare Google Analytics data

The following tutorial illustrates how the raw data for the Big-P example in Chapter 2 was collected and prepared via Google BigQuery and R. Before we get started, note an important aspect of a data warehouse solution like BigQuery in contrast to common applications of RDBS. As data warehouses are used in a more flexible way than relational databases, it is not uncommon to store data files/tables containing the same variables separately in various tables, for example to store one table per day or year of a panel dataset. On Google BigQuery, this partitioning of datasets into several components can additionally make sense for cost reasons. Suppose you want to only compute summary statistics for certain variables over a given time frame. If all observations of a large dataset are stored in one standard BigQuery table, such a query results in processing GBs or TBs of data, as the observations from the corresponding time frame need to be filtered out of the entire dataset. Partitioning the data into several subsets helps avoid this, as BigQuery has several features that allow the definition of SQL queries to be run on partitioned data. The publicly available Google Analytics dataset is organized in such a partitioned way. The data is stored in several tables (one for each day of the observation period), whereby the last few characters of the table name contain the date of the corresponding observation day (such as the one used in the example above: `ga_sessions_20170801`). If we want to combine data from several of those tables, we can use the wildcard character (`*`) to indicate that the BigQuery should consider all tables matching the table name up to the `*`: `FROM bigquery-public-data.google_analytics_sample.ga_sessions_*`.

We proceed by first connecting the R session with GoogleBigQuery.

```
# fix vars
# the project ID on BigQuery (billing must be enabled)
BILLING <- "YOUR-BILLING-PROJECT-ID"
# the project name on BigQuery
PROJECT <- "bigquery-public-data"
DATASET <- "google_analytics_sample"

# connect to DB on BigQuery
con <- dbConnect(
    bigrquery::bigquery(),
    project = PROJECT,
```

```
        dataset = DATASET,
        billing = BILLING
)
```

The query combines all Google Analytics data recorded from the beginning of 2016 to the end of 2017 via WHERE _TABLE_SUFFIX BETWEEN '20160101' AND '20171231'. This gives us all the raw data used in the Big-P analysis shown in Chapter 2.

```
# run query
query <-
"
SELECT
totals.visits,
totals.transactions,
trafficSource.source,
device.browser,
device.isMobile,
geoNetwork.city,
geoNetwork.country,
channelGrouping
FROM `bigquery-public-data.google_analytics_sample.ga_sessions_*`
WHERE _TABLE_SUFFIX BETWEEN '20160101' AND '20171231';
"
ga <- as.data.table(dbGetQuery(con, query, page_size=15000))
```

Finally, we use data.table and basic R to prepare the final analytic dataset and write it on disk.

```
# further cleaning and coding via data.table and basic R
ga$transactions[is.na(ga$transactions)] <- 0
ga <- ga[ga$city!="not available in demo dataset",]
ga$purchase <- as.integer(0<ga$transactions)
ga$transactions <- NULL
ga_p <- ga[purchase==1]
ga_rest <- ga[purchase==0][sample(1:nrow(ga[purchase==0]), 45000)]
ga <- rbindlist(list(ga_p, ga_rest))
potential_sources <- table(ga$source)
potential_sources <- names(potential_sources[1<potential_sources])
ga <- ga[ga$source %in% potential_sources,]
```

```
# store dataset on local hard disk
fwrite(ga, file="data/ga.csv")

# clean up
dbDisconnect(con)
```

Note how we combine BigQuery as our data warehouse with basic R for data preparation. Solutions like BigQuery are particularly useful for this kind of approach as part of an analytics project: Large operations such as the selection of columns/-variables from large-scale data sources are handled within the warehouse in the cloud, and the refinement/cleaning steps can then be implemented locally on a much smaller subset.[13]

Note the wildcard character (*) in the query is used to fetch data from several partitions of the overall dataset.

8.8 Data lakes and simple storage service

Broadly speaking a data lake is where all your data resides (these days, this is typically somewhere in the cloud). The data is simply stored in whatever file format and in simple terms organized in folders and sub-folders. In the same data lake you might thus store CSV files, SQL database dumps, log files, image files, raw text, etc. In addition, you typically have many options to define access rights to files, including to easily make them accessible for download to the public. For a simple data analytics project in the context of economic research or business analytics, the data lake in the cloud concept is a useful tool to store all project-related raw data files. On the one hand you avoid running into troubles with occupying gigabytes or terabytes of your local hard disk with files that are relevant but only rarely imported/worked with. On the other hand you can properly organize all the raw data for reproducibility purposes and easily share the files with colleagues (and eventually the public). For example, you can use one main folder (one "bucket") for an entire analytics project, store all the raw data in one sub-folder (for reproduction purposes), and store all the final analytic datasets in another sub-folder for replication purposes and more frequent access as well as sharing across a team of co-workers.

---

[13] Since in this example the queried subset is not particularly large, it is easier to perform the data preparation locally in R. However, in other situations it might make sense to use SQL in BigQuery more extensively for data preparation tasks that would require a lot of RAM.

There are several types of cloud-based data lake solutions available, many of which are primarily focused on corporate data storage and provide a variety of services (for example, AWS Lake Formation or Azure Data Lake) that might go well beyond the data analytics perspective taken in this book. However, most of these solutions build in the end on a so-called simple storage service such as AWS S3 or Google Cloud Storage, which build the core of the lake – the place where the data is actually stored and accessed. In the following, we will look at how to use such a simple storage service (AWS S3) as a data lake in simple analytics projects.[14]

Finally, we will look at a very interesting approach to combine the concept of a data lake with the concept of a data warehouse. That is, we briefly look at solutions of how some analytics tools (specifically, a tool called Amazon Athena) can directly be used to query/analyze the data stored in the simple storage service.

8.8.1   AWS S3 with R: First steps

For the following first steps with AWS S3 and R, you will need an AWS account (the same as above for EC2) and IAM credentials from your AWS account with the right to access S3.[15] Finally, you will have to install the `aws.s3` package in R in order to access S3 via R: `install.packages("aws.s3")`.

To initiate an R session in which you connect to S3, `aws.s3` (Leeper, 2020) must be loaded and the following environment variables must be set:

- `AWS_ACCESS_KEY_ID`: your access key ID (of the keypair with rights to use S3)
- `AWS_SECRET_KEY`: your access key (of the keypair with rights to use S3)
- `REGION`: the region in which your S3 buckets are/will be located (e.g., `"eu-central-1"`)

```
# load packages
library(aws.s3)


# set environment variables with your AWS S3 credentials
Sys.setenv("AWS_ACCESS_KEY_ID" = AWS_ACCESS_KEY_ID,
           "AWS_SECRET_ACCESS_KEY" = AWS_SECRET_KEY,
           "AWS_DEFAULT_REGION" = REGION)
```

---

[14] Essentially, all other major cloud computing providers offer very similar services with very similar features to AWS S3. Conceptually, you could thus easily use one of these other services. The examples here focus on AWS S3 primarily for simplicity (as we have already set up AWS credentials etc.), and the straightforward way to connect to AWS S3 via R.

[15] There are many ways to create these credentials, and many ideas about which ones to use. A very simple and reasonable instruction how to do this can be found here: https://binaryguy.tech/a ws/s3/create-iam-user-to-access-s3/.

In a first step, we create a project bucket (the main repository for our project) to store all the data of our analytics project. All the raw data can be placed directly in this main folder. Then, we add one sub-folder to this bucket: `analytic_data` (for the cleaned/prepared datasets underlying the analyses in the project).[16]

```r
# fix variable for bucket name
BUCKET <- "tlc-trips"
# create project bucket
put_bucket(BUCKET)
# create folders
put_folder("raw_data", BUCKET)
put_folder("analytic_data", BUCKET)
```

8.8.2 Uploading data to S3

Now we can start uploading the data to the bucket (and the sub-folder). For example, to remain within the context of the TLC taxi trips data, we upload the original Parquet files directly to the bucket and the prepared CSV file to `analytic_data`. For large files (larger than 100MB) it is recommended to use the multipart option (upload of file in several parts; `multipart=TRUE`).

```r
# upload to bucket
# final analytic dataset
put_object(
  file = "data/tlc_trips.csv", # the file you want to upload
  object = "analytic_data/tlc_trips.csv", # name of the file in the bucket
  bucket = BUCKET,
  multipart = TRUE
)

# upload raw data
file_paths <- list.files("data/tlc_trips/raw_data", full.names = TRUE)
lapply(file_paths,
       put_object,
       bucket=BUCKET,
       multipart=TRUE)
```

---

[16]Note that, technically, the explicit creation of folders is not necessary, as S3 uses slashes (/) in file names on S3 to make them appear to be in a particular folder. However, when using the AWS S3 console in the browser, defining folders explicitly can make more sense from the user's perspective.

8.8.3 More than just simple storage: S3 + Amazon Athena

There are several implementations of interfaces with Amazon Athena in R. Here, we will rely on `AWR.Athena` (Fultz and Daróczi, 2019) (run `install.packages("AWR.Athena")`), which allows interacting with Amazon Athena via the familiar DBI package (R Special Interest Group on Databases (R-SIG-DB) et al., 2022).

```
# SET UP ----------------------


# load packages
library(DBI)
library(aws.s3)
# aws credentials with Athena and S3 rights and region
AWS_ACCESS_KEY_ID <- "YOUR_KEY_ID"
AWS_ACCESS_KEY <- "YOUR_KEY"
REGION <- "eu-central-1"



# establish AWS connection
Sys.setenv("AWS_ACCESS_KEY_ID" = AWS_ACCESS_KEY_ID,
           "AWS_SECRET_ACCESS_KEY" = AWS_ACCESS_KEY,
           "AWS_DEFAULT_REGION" = REGION)
```

Create a bucket for the output.

```
OUTPUT_BUCKET <- "bda-athena"
put_bucket(OUTPUT_BUCKET, region="us-east-1")
```

Now we can connect to Amazon Athena to query data from files in S3 via the RJDBC package (Urbanek, 2022).

```
# load packages
library(RJDBC)
library(DBI)


# download Athena JDBC driver
URL <- "https://s3.amazonaws.com/athena-downloads/drivers/JDBC/"
VERSION <- "AthenaJDBC_1.1.0/AthenaJDBC41-1.1.0.jar"
DRV_FILE <- "AthenaJDBC41-1.1.0.jar"
download.file(paste0(URL, VERSION), destfile = DRV_FILE)
```

```
# connect to JDBC
athena <- JDBC(driverClass = "com.amazonaws.athena.jdbc.AthenaDriver",
  DRV_FILE, identifier.quote = "'")
# connect to Athena
con <- dbConnect(athena, "jdbc:awsathena://athena.us-east-1.amazonaws.com:443/",
  s3_staging_dir = "s3://bda-athena", user = AWS_ACCESS_KEY_ID,
  password = AWS_ACCESS_KEY)
```

In order to query data stored in S3 via Amazon Athena, we need to create an *external table* in Athena, which will be based on data stored in S3.

```
query_create_table <-
"
CREATE EXTERNAL TABLE default.trips (
  `vendor_name` string,
  `Trip_Pickup_DateTime` string,
  `Trip_Dropoff_DateTime` string,
  `Passenger_Count` int,
  `Trip_Distance` double,
  `Start_Lon` double,
  `Start_Lat` double,
  `Rate_Code` string,
  `store_and_forward` string,
  `End_Lon` double,
  `End_Lat` double,
  `Payment_Type` string,
  `Fare_Amt` double,
  `surcharge` double,
  `mta_tax` string,
  `Tip_Amt` double,
  `Tolls_Amt` double,
  `Total_Amt` double
)
ROW FORMAT DELIMITED
FIELDS TERMINATED BY ','
STORED AS TEXTFILE
LOCATION 's3://tlc-trips/analytic_data/'
"
dbSendQuery(con, query_create_table)
```

Run a test query to verify the table.

```
test_query <-
"
SELECT *
FROM default.trips
LIMIT 10
"

test <- dbGetQuery(con, test_query)
dim(test)
```

```
## [1] 10 18
```

Finally, close the connection.

```
dbDisconnect(con)
```

```
## [1] TRUE
```

8.9   Wrapping up

- It is good practice to set up all of the high-level *pipeline* in the same language (here R). This substantially facilitates your workflow and makes your overall pipeline easier to maintain. Importantly, as illustrated in the sections above, this practice does not mean that all of the underlying data processing is actually done in R. We simply use R as the highest-level layer and call a range of services under the hood to handle each of the pipeline components as efficiently as possible.
- *Apache Arrow* allows you to combine and correct raw data without exceeding RAM; in addition it facilitates working with newer (big) data formats for columnar data storage systems (like *Apache Parquet*).
- *RDBMSs* such as *SQLite* or *MySQL* and analytics databases such as *Druid* help you store and organize clean/structured data for analytics purposes locally or in the cloud.
- *RDBMSs* like SQLite are *row-based* (changing a value means changing a row), while modern analytics databases are usually *column*-based (changing a value means modifying one column).
- Row-based databases are recommended when your analytics workflow includes a lot of tables, table joins, and frequent filtering for specific observations with variables from several tables. Column-based databases are recommended for analytics workflows involving less frequent but large-scale data aggregation tasks.
- *Data warehouse* solutions like *Google BigQuery* are useful to store and query large

(semi-)structured datasets but are more flexible regarding hierarchical data and file formats than traditional RDBMSs.

- *Data lakes* and simple storage services are the all-purpose tools to store vast amounts of data in any format in the cloud. Typically, solutions like *AWS S3* are a great option to store all of the raw data related to a data analytics project.

9

Big Data Cleaning and Transformation

Preceding the filtering/selection/aggregation of raw data, data cleaning and transformation typically have to be run on large volumes of raw data before the observations and variables of interest can be further analyzed. Typical data cleaning tasks involve:

- Normalization/standardization (across entities, categories, observation periods).
- Coding of additional variables (indicators, strings to categorical, etc.).
- Removing/adding covariates.
- Merging/joining datasets.
- Properly defining data types for each variable.

All of these steps are very common tasks when working with data for analytics purposes, independent of the size of the dataset. However, as most of the techniques and software developed for such tasks is meant to process data in memory, performing these tasks on large datasets can be challenging. Data cleaning workflows you are perfectly familiar with might slow down substantially or crash due to a lack of memory (RAM), particularly if the data preparation step involves merging/joining two datasets. Other potential bottlenecks are the parsing of large files (CPU) or intensive reading from and writing to the hard disk (Mass storage).

In practice, the most critical bottleneck of common data preparation tasks is often a lack of RAM. In the following, we thus explore two strategies that broadly build on the idea of *virtual memory* (using parts of the hard disk as RAM) and/or *lazy evaluation* (only loading/processing the part of a dataset really required).

9.1 Out-of-memory strategies and lazy evaluation: Practical basics

Virtual memory is in simple words an approach to combining the RAM and mass storage components in order to cope with a lack of RAM. Modern operating systems come with a virtual memory manager that automatically handles the swapping between RAM and the hard-disk, when running processes that use up too much RAM. However, a virtual memory manager is not specifically developed to perform this task in the context of data analysis. Several strategies have thus been developed to build on the basic idea of *virtual memory* in the context of data analysis tasks.

- *Chunked data files on disk*: The data analytics software 'partitions' the dataset, and maps and stores the chunks of raw data on disk. What is actually 'read' into RAM when importing the data file with this approach is the mapping to the partitions of the actual dataset (the data structure) and some metadata describing the dataset. In R, this approach is implemented in the `ff` package (Adler et al., 2022) and several packages building on `ff`. In this approach, the usage of disk space and the linking between RAM and files on disk is very explicit (and clearly visible to the user).

- *Memory mapped files and shared memory*: The data analytics software uses segments of virtual memory for the dataset and allows different programs/processes to access it in the same memory segment. Thus, virtual memory is explicitly allocated for one or several specific data analytics tasks. In R, this approach is notably implemented in the `bigmemory` package (Kane et al., 2013) and several packages building on `bigmemory`.

A conceptually related but differently focused approach is the *lazy evaluation* implemented in Apache Arrow and the corresponding `arrow` package (Richardson et al., 2022). While Apache Arrow is basically a platform for in-memory columnar data, it is optimized for processing large amounts of data and working with datasets that actually do not fit into memory. The way this is done is that instructions on what to do with a dataset are not evaluated step-by-step on the spot but all together at the point of actually loading the data into R. That is, we can connect to a dataset via `arrow`, see its variables, etc., give instructions of which observations to filter out and which columns to select, all before we read the dataset into RAM. In comparison to the strategies outlined above, this approach is usually much faster but might still lead to a situation with a lack of memory.

In the following subsections we briefly look at how to set up an R session for data preparation purposes with any of these approaches (`ff`, `bigmemory`, `arrow`) and look at some of the conceptual basics behind the approaches.

9.1.1   Chunking data with the `ff` package

We first install and load the `ff` and `ffbase` (de Jonge et al., 2023) packages, as well as the `pryr` package. We use the familiar `flights.csv` dataset[1] For the sake of the example, we only use a fraction of the original dataset. On disk, the dataset is about 30MB:

---

[1]Data from the same source is also used in the code examples given in Kane et al. (2013).

```
fs::file_size("data/flights.csv")
```

```
## 29.5M
```

However, loading the entire dataset of several GBs would work just fine, using the ff-approach.

When importing data via the ff package, we first have to set up a directory where ff can store the partitioned dataset (recall that this is explicitly/visibly done on disk). We call this new directory ff_files.

```
# SET UP --------------
# install.packages(c("ff", "ffbase"))
# you might have to install the ffbase package directly from GitHub:
# devtools::install_github("edwindj/ffbase", subdir="pkg")
# load packages
library(ff)
library(ffbase)
library(data.table) # for comparison


# create directory for ff chunks, and assign directory to ff
system("mkdir ff_files")
options(fftempdir = "ff_files")
```

Now we can read in the data with read.table.ffdf. In order to better understand the underlying concept, we also import the data into a common data.table object via fread() and then look at the size of the objects resulting from the two 'import' approaches in the R environment with object.size().

```
# usual in-memory csv import
flights_dt <- fread("data/flights.csv")


# out-of-memory approach
flights <-
    read.table.ffdf(file="data/flights.csv",
                    sep=",",
                    VERBOSE=TRUE,
                    header=TRUE,
                    next.rows=100000,
                    colClasses=NA)
```

```
## read.table.ffdf 1..100000 (100000)  csv-read=0.433sec ffdf-write=0.076sec
## read.table.ffdf 100001..200000 (100000)  csv-read=0.415sec ffdf-write=0.045sec
## read.table.ffdf 200001..300000 (100000)  csv-read=0.442sec ffdf-write=0.056sec
## read.table.ffdf 300001..336776 (36776)  csv-read=0.169sec ffdf-write=0.031sec
##  csv-read=1.459sec  ffdf-write=0.208sec  TOTAL=1.667sec
```

```
# compare object sizes
object.size(flights) # out-of-memory approach
```

```
## 949976 bytes
```

```
object.size(flights_dt) # common data.table
```

```
## 32569024 bytes
```

Note that there are two substantial differences to what we have previously seen when using fread(). It takes much longer to import a CSV into the ff_files structure. However, the RAM allocated to it is much smaller. This is exactly what we would expect, keeping in mind what read.table.ffdf() does in comparison to what fread() does. Now we can actually have a look at the data chunks created by ff.

```
# show the files in the directory keeping the chunks
head(list.files("ff_files"))
```

```
## [1] "ffdf991b190c435c.ff" "ffdf991b1ad3a49c.ff"
## [3] "ffdf991b1ba739af.ff" "ffdf991b21ef4115.ff"
## [5] "ffdf991b2464b9c4.ff" "ffdf991b252df4ac.ff"
```

9.1.2 Memory mapping with bigmemory

The bigmemory package handles data in matrices and therefore only accepts data values of identical data type. Before importing data via the bigmemory package, we thus have to ensure that all variables in the raw data can be imported in a common type.

```
# SET UP ----------------
```

```
# load packages
library(bigmemory)
library(biganalytics)
```

```
# import the data
flights <- read.big.matrix("data/flights.csv",
                           type="integer",
                           header=TRUE,
                           backingfile="flights.bin",
                           descriptorfile="flights.desc")
```

Note that, similar to the `ff` example, `read.big.matrix()` creates a local file `flights.bin` on disk that is linked to the `flights` object in RAM. From looking at the imported file, we see that various variable values have been discarded. This is because we have forced all variables to be of type `"integer"` when importing the dataset.

```
object.size(flights)
```

```
## 696 bytes
```

```
str(flights)
```

```
## Formal class 'big.matrix' [package "bigmemory"] with 1
slot
## ..@ address:<externalptr>
```

Again, the object representing the dataset in R does not contain the actual data (it does not even take up a KB of memory).

9.1.3 Connecting to Apache Arrow

```
# SET UP ----------------

# load packages
library(arrow)

# import the data
flights <- read_csv_arrow("data/flights.csv",
                          as_data_frame = FALSE)
```

Note the `as_data_frame=FALSE` in the function call. This instructs Arrow to establish a connection to the file and read some of the data (to understand what is in the file), but not actually import the whole CSV.

```
summary(flights)
```

```
##               Length Class        Mode
## year          336776 ChunkedArray environment
## month         336776 ChunkedArray environment
## day           336776 ChunkedArray environment
## dep_time      336776 ChunkedArray environment
## sched_dep_time 336776 ChunkedArray environment
## dep_delay     336776 ChunkedArray environment
## arr_time      336776 ChunkedArray environment
## sched_arr_time 336776 ChunkedArray environment
## arr_delay     336776 ChunkedArray environment
## carrier       336776 ChunkedArray environment
## flight        336776 ChunkedArray environment
## tailnum       336776 ChunkedArray environment
## origin        336776 ChunkedArray environment
## dest          336776 ChunkedArray environment
## air_time      336776 ChunkedArray environment
## distance      336776 ChunkedArray environment
## hour          336776 ChunkedArray environment
## minute        336776 ChunkedArray environment
## time_hour     336776 ChunkedArray environment
```

```
object.size(flights)
```

```
## 488 bytes
```

Again, we notice that the flights object is much smaller than the actual dataset on disk.

9.2 Big Data preparation tutorial with ff

9.2.1 Set up

The following code and data examples build on Walkowiak (2016), Chapter 3.[2] The set up for our analysis script involves the loading of the ff and ffbase packages, the initialization of fixed variables to hold the paths to the datasets, and the

---

[2]You can download the original datasets used in these examples from https://github.com/PacktPublishing/Big-Data-Analytics-with-R/tree/master/Chapter%203.

creation and assignment of a new local directory `ff_files` in which the binary flat file-partitioned chunks of the original datasets will be stored.

```
## SET UP ------------------------

# create and set directory for ff files
system("mkdir ff_files")
options(fftempdir = "ff_files")

# load packages
library(ff)
library(ffbase)
library(pryr)

# fix vars
FLIGHTS_DATA <- "data/flights_sep_oct15.txt"
AIRLINES_DATA <- "data/airline_id.csv"
```

9.2.2 Data import

In a first step we read (or 'upload') the data into R. This step involves the creation of the binary chunked files as well as the mapping of these files and the metadata. In comparison to the traditional `read.csv` approach, you will notice two things. On the one hand the data import takes longer; on the other hand it uses up much less RAM than with `read.csv`.

```
# DATA IMPORT -----------------

# check memory used
mem_used()


## 1.79 GB


# 1. Upload flights_sep_oct15.txt and airline_id.csv files from flat files.

system.time(flights.ff <- read.table.ffdf(file=FLIGHTS_DATA,
                                          sep=",",
                                          VERBOSE=TRUE,
                                          header=TRUE,
                                          next.rows=100000,
                                          colClasses=NA))
```

```
## read.table.ffdf 1..100000 (100000)  csv-read=0.589sec ffdf-write=0.091sec
## read.table.ffdf 100001..200000 (100000)  csv-read=0.587sec ffdf-write=0.065sec
## read.table.ffdf 200001..300000 (100000)  csv-read=0.581sec ffdf-write=0.063sec
## read.table.ffdf 300001..400000 (100000)  csv-read=0.603sec ffdf-write=0.074sec
## read.table.ffdf 400001..500000 (100000)  csv-read=0.657sec ffdf-write=0.074sec
## read.table.ffdf 500001..600000 (100000)  csv-read=0.675sec ffdf-write=0.073sec
## read.table.ffdf 600001..700000 (100000)  csv-read=0.613sec ffdf-write=0.076sec
## read.table.ffdf 700001..800000 (100000)  csv-read=0.648sec ffdf-write=0.089sec
## read.table.ffdf 800001..900000 (100000)  csv-read=0.635sec ffdf-write=0.07sec
## read.table.ffdf 900001..951111 (51111)  csv-read=0.331sec ffdf-write=0.063sec
##  csv-read=5.919sec  ffdf-write=0.738sec  TOTAL=6.657sec

##    user  system elapsed
##   5.733   0.821   6.669


system.time(airlines.ff <- read.csv.ffdf(file= AIRLINES_DATA,
                              VERBOSE=TRUE,
                              header=TRUE,
                              next.rows=100000,
                              colClasses=NA))


## read.table.ffdf 1..1607 (1607)  csv-read=0.005sec ffdf-write=0.005sec
##  csv-read=0.005sec  ffdf-write=0.005sec  TOTAL=0.01sec

##    user  system elapsed
##   0.010   0.000   0.019


# check memory used
mem_used()


## 1.79 GB
```

Comparison with read.table

```
# Using read.table()
system.time(flights.table <- read.table(FLIGHTS_DATA,
                                  sep=",",
                                  header=TRUE))


##    user  system elapsed
##   6.527   0.526   7.089
```

```
system.time(airlines.table <- read.csv(AIRLINES_DATA,
                                        header = TRUE))
```

```
##    user  system elapsed
##   0.002   0.000   0.002
```

```
# check the memory used
mem_used()
```

```
## 1.93 GB
```

9.2.3   Inspect imported files

A particularly useful aspect of working with the ff package and the packages building on it is that many of the simple R functions that work on normal data.frames in RAM also work on ff_files files. Hence, without actually having loaded the entire raw data of a large dataset into RAM, we can quickly get an overview of the key characteristics, such as the number of observations and the number of variables.

```
# 2. Inspect the ff_files objects.
## For flights.ff object:
class(flights.ff)
```

```
## [1] "ffdf"
```

```
dim(flights.ff)
```

```
## [1] 951111      28
```

```
## For airlines.ff object:
class(airlines.ff)
```

```
## [1] "ffdf"
```

```
dim(airlines.ff)
```

```
## [1] 1607     2
```

9.2.4 Data cleaning and transformation

After inspecting the data, we go through several steps of cleaning and transformation, with the goal of then merging the two datasets. That is, we want to create a new dataset that contains detailed flight information but with additional information on the carriers/airlines. First, we want to rename some of the variables.

```
# step 1:
# Rename "Code" variable from airlines.ff
# to "AIRLINE_ID" and "Description" into "AIRLINE_NM".
names(airlines.ff) <- c("AIRLINE_ID", "AIRLINE_NM")
names(airlines.ff)
```

```
## [1] "AIRLINE_ID" "AIRLINE_NM"
```

```
str(airlines.ff[1:20,])
```

```
## 'data.frame': 20 obs. of 2 variables:
## $ AIRLINE_ID: int 19031 19032 19033 19034 19035 19036
19037 19038 19039 19040 ...
## $ AIRLINE_NM: Factor w/ 1607 levels "40-Mile Air:
Q5",..: 945 1025 503 721 64 725 1194 99 1395 276 ...
```

Now we can join the two datasets via the unique airline identifier `"AIRLINE_ID"`. Note that these kinds of operations would usually take up substantially more RAM on the spot, if both original datasets were also fully loaded into RAM. As illustrated by the `mem_change()` function, this is not the case here. All that is needed is a small chunk of RAM to keep the metadata and mapping-information of the new `ff_files` object; all the actual data is cached on the hard disk.

```
# merge of ff_files objects
mem_change(flights.data.ff <- merge.ffdf(flights.ff,
                                         airlines.ff,
                                         by="AIRLINE_ID"))
```

```
## 774 kB
```

```
#The new object is only 551.2 KB in size
class(flights.data.ff)
```

```
## [1] "ffdf"
```

```
dim(flights.data.ff)
```

```
## [1] 951111     29
```

```
names(flights.data.ff)
```

```
##  [1] "YEAR"              "MONTH"
##  [3] "DAY_OF_MONTH"      "DAY_OF_WEEK"
##  [5] "FL_DATE"           "UNIQUE_CARRIER"
##  [7] "AIRLINE_ID"        "TAIL_NUM"
##  [9] "FL_NUM"            "ORIGIN_AIRPORT_ID"
## [11] "ORIGIN"            "ORIGIN_CITY_NAME"
## [13] "ORIGIN_STATE_NM"   "ORIGIN_WAC"
## [15] "DEST_AIRPORT_ID"   "DEST"
## [17] "DEST_CITY_NAME"    "DEST_STATE_NM"
## [19] "DEST_WAC"          "DEP_TIME"
## [21] "DEP_DELAY"         "ARR_TIME"
## [23] "ARR_DELAY"         "CANCELLED"
## [25] "CANCELLATION_CODE" "DIVERTED"
## [27] "AIR_TIME"          "DISTANCE"
## [29] "AIRLINE_NM"
```

9.2.5   Inspect difference in in-memory operation

In comparison to the ff-approach, performing the merge in memory needs more resources:

```
##For flights.table:
names(airlines.table) <- c("AIRLINE_ID", "AIRLINE_NM")
names(airlines.table)
```

```
## [1] "AIRLINE_ID" "AIRLINE_NM"
```

```
str(airlines.table[1:20,])
```

```
## 'data.frame': 20 obs. of 2 variables:
## $ AIRLINE_ID: int 19031 19032 19033 19034 19035 19036
19037 19038 19039 19040 ...
## $ AIRLINE_NM: chr "Mackey International Inc.: MAC" "Munz
Northern Airlines Inc.: XY" "Cochise Airlines Inc.: COC"
"Golden Gate Airlines Inc.: GSA" ...
```

```
# check memory usage of merge in RAM
mem_change(flights.data.table <- merge(flights.table,
                                       airlines.table,
                                       by="AIRLINE_ID"))
```

```
## 161 MB
```

```
#The new object is already 105.7 MB in size
#A rapid spike in RAM use when processing
```

9.2.6 Subsetting

Now, we want to filter out some observations as well as select only specific variables for a subset of the overall dataset.

```
mem_used()
```

```
## 2.09 GB
```

```
# Subset the ff_files object flights.data.ff:
subs1.ff <-
     subset.ffdf(flights.data.ff,
                 CANCELLED == 1,
                 select = c(FL_DATE,
                            AIRLINE_ID,
                            ORIGIN_CITY_NAME,
                            ORIGIN_STATE_NM,
                            DEST_CITY_NAME,
                            DEST_STATE_NM,
                            CANCELLATION_CODE))
```

```
dim(subs1.ff)
```

```
## [1] 4529      7
```

```
mem_used()
```

```
## 2.09 GB
```

9.2.7   Save/load/export ff files

In order to better organize and easily reload the newly created ff_files files, we can explicitly save them to disk.

```
# Save a newly created ff_files object to a data file:
# (7 files (one for each column) created in the ffdb directory)
save.ffdf(subs1.ff, overwrite = TRUE)
```

If we want to reload a previously saved ff_files object, we do not have to go through the chunking of the raw data file again but can very quickly load the data mapping and metadata into RAM in order to further work with the data (stored on disk).

```
# Loading previously saved ff_files files:
rm(subs1.ff)
#gc()
load.ffdf("ffdb")
# check the class and structure of the loaded data
class(subs1.ff)
```

```
## [1] "ffdf"
```

```
dim(subs1.ff)
```

```
## [1] 4529    7
```

```
dimnames(subs1.ff)
```

```
## [[1]]
## NULL
##
## [[2]]
## [1] "FL_DATE"           "AIRLINE_ID"
## [3] "ORIGIN_CITY_NAME"  "ORIGIN_STATE_NM"
## [5] "DEST_CITY_NAME"    "DEST_STATE_NM"
## [7] "CANCELLATION_CODE"
```

If we want to store an ff_files dataset in a format more accessible for other users (such as CSV), we can do so as follows. This last step is also quite common in practice. The initial raw dataset is very large; thus we perform all the theoretically very memory-intensive tasks of preparing the analytic dataset via ff and then store the (often much smaller) analytic dataset in a more accessible CSV file in order to later read it into RAM and run more computationally intensive analyses directly in RAM.

```
#  Export subs1.ff into CSV and TXT files:
write.csv.ffdf(subs1.ff, "subset1.csv")
```

9.3   Big Data preparation tutorial with `arrow`

We begin by initializing our R session as in the short `arrow` introduction above.

```
# SET UP ----------------

# load packages
library(arrow)
library(dplyr)
library(pryr) # for profiling

# fix vars
FLIGHTS_DATA <- "data/flights_sep_oct15.txt"
AIRLINES_DATA <- "data/airline_id.csv"

# import the data
flights <- read_csv_arrow(FLIGHTS_DATA,
                   as_data_frame = FALSE)
airlines <- read_csv_arrow(AIRLINES_DATA,
                   as_data_frame = FALSE)
```

Note how the data from the CSV files is not actually read into RAM yet. The created objects `flights` and `airlines` are not data frames (yet) and occupy hardly any RAM.

```
class(flights)
```

```
## [1] "Table"        "ArrowTabular" "ArrowObject"
## [4] "R6"
```

```
class(airlines)
```

```
## [1] "Table"        "ArrowTabular" "ArrowObject"
```

```
## [4] "R6"
```

```
object_size(flights)
```

```
## 283.62 kB
```

```
object_size(airlines)
```

```
## 283.62 kB
```

In analogy to the ff tutorial above, we go through the same data preparation steps. First, we rename the variables in airlines to ensure that the variable names are consistent with the flights data frame.

```
# step 1:
# Rename "Code" variable from airlines.ff to "AIRLINE_ID"
# and "Description" into "AIRLINE_NM".
names(airlines) <- c("AIRLINE_ID", "AIRLINE_NM")
names(airlines)
```

```
## [1] "AIRLINE_ID" "AIRLINE_NM"
```

In a second step, the two data frames are merged/joined. The arrow package follows dplyr-syntax regarding data preparation tasks. That is, we can directly build on functions like

```
# merge the two datasets via Arrow
flights.data.ar <- inner_join(airlines, flights, by="AIRLINE_ID")
object_size(flights.data.ar)
```

```
## 647.74 kB
```

In a last step, we filter the resulting dataset for cancelled flights and select only some of the available variables.

Now, we want to filter out some observations as well as select only specific variables for a subset of the overall dataset. As Arrow works with the dplyr back-end, we can directly use the typical dplyr-syntax to combine selection of columns and filtering of rows.

```
# Subset the ff_files object flights.data.ff:
subs1.ar <-
```

```
flights.data.ar %>%
filter(CANCELLED == 1) %>%
select(FL_DATE,
       AIRLINE_ID,
       ORIGIN_CITY_NAME,
       ORIGIN_STATE_NM,
       DEST_CITY_NAME,
       DEST_STATE_NM,
       CANCELLATION_CODE)
```

```
object_size(subs1.ar)
```

```
## 591.21 kB
```

Again, this operation hardly affected RAM usage by R. Note, though, that in contrast to the ff-approach, Arrow has actually not yet created the new subset subs1.ar. In fact, it has not even really imported the data or merged the two datasets. This is the effect of the lazy evaluation approach implemented in arrow. To further process the data in subs1.ar with other functions (outside of arrow), we need to actually trigger the evaluation of all the data preparation steps we have just instructed R to do. This is done via collect().

```
mem_change(subs1.ar.df <- collect(subs1.ar))
```

```
## 2.47 MB
```

```
class(subs1.ar.df)
```

```
## [1] "tbl_df"      "tbl"           "data.frame"
```

```
object_size(subs1.ar.df)
```

```
## 57.15 kB
```

Note how in this tutorial, the final subset is substantially smaller than the initial two datasets. Hence, in this case it is fine to actually load this into RAM as a data frame. However, this is not a necessary part of the workflow. Instead of calling collect(), you can then trigger the computation of all the data preparation steps via compute() and, for example, store the resulting arrow table to a CSV file.

```
subs1.ar %>%
        compute() %>%
        write_csv_arrow(file="data/subs1.ar.csv")
```

9.4 Wrapping up

- Typically, the raw/uncleaned data is the critical bottleneck in terms of data volume, particularly as the selection and filtering of the overall dataset in the preparation of analytic datasets can only work properly with cleaned data.
- *Out-of-memory* strategies are based on the concept of virtual memory and are key to cleaning large amounts of data locally.
- The `ff` *package* provides a high-level R interface to an out-of-memory approach. Most functions in `ff` and the corresponding `ffbase` package come with a syntax very similar to the basic R syntax for data cleaning and manipulation.
- The basic idea behind `ff` is to store the data in chunked format in an easily accessible way on the hard disk and only keep the metadata of a dataset (e.g., variable names) in an R object in RAM while working on the dataset.
- The `arrow` package offers similar functionality based on a slightly different approach called *lazy evaluation* (only evaluate data manipulation/cleaning tasks once the data is pulled into R). Unlike `ff`, `arrow` closely follows the `dplyr` syntax rather than basic R syntax for data cleaning tasks.

10

Descriptive Statistics and Aggregation

10.1 Data aggregation: The 'split-apply-combine' strategy

The 'split-apply-combine' strategy plays an important role in many data analysis tasks, ranging from data preparation to summary statistics and model-fitting.[1] The strategy can be defined as "break up a problem into manageable pieces, operate on each piece independently, and then put all the pieces back together." (Wickham, 2011, p. 1)

Many R users are familiar with the basic concept of split-apply-combine implemented in the `plyr` package intended for normal in-memory operations (dataset fits into RAM). Here, we explore the options for split-apply-combine approaches to large datasets that do not fit into RAM.

10.2 Data aggregation with chunked data files

In this tutorial we explore the world of New York's famous Yellow Cabs. In a first step, we will focus on the `ff`-based approach to employ parts of the hard disk as 'virtual memory'. This means, that all of the examples are easily scalable without risking too much memory pressure. Given the size of the entire TLC database (over 200GB), we will only use one million taxi trip records.[2]

Data import

First, we read the raw taxi trip records into R with the `ff` package.

```
# load packages
library(ff)
```

---

[1]Moreover, 'split-apply-combine' is closely related to a core strategy of Big Data Analytics with distributed systems (MapReduce).

[2]Note that the code examples below could also be run based on the entire TLC database (provided that there is enough hard-disk space available). But, creating the `ff` chunked file structure for a 200GB CSV would take hours or even days.

```
library(ffbase)

# set up the ff directory (for data file chunks)
if (!dir.exists("fftaxi")){
    system("mkdir fftaxi")
}
options(fftempdir = "fftaxi")

# import the first one million observations
taxi <- read.table.ffdf(file = "data/tlc_trips.csv",
                        sep = ",",
                        header = TRUE,
                        next.rows = 100000,
                        # colClasses= col_classes,
                        nrows = 1000000
                        )
```

Following the data documentation provided by TLC, we give the columns of our dataset more meaningful names and remove the empty columns (some covariates are only collected in later years).

When inspecting the factor variables of the dataset, we notice that some of the values are not standardized/normalized, and the resulting factor levels are, therefore, somewhat ambiguous. We should clean this before getting into data aggregation tasks. Note the ff-specific syntax needed to recode the factor.

```
# inspect the factor levels
levels(taxi$Payment_Type)
```

```
## [1] "Cash"      "CASH"      "Credit"    "CREDIT"
## [5] "Dispute"   "No Charge"
```

```
# recode them
levels(taxi$Payment_Type) <- tolower(levels(taxi$Payment_Type))
taxi$Payment_Type <- ff(taxi$Payment_Type,
                    levels = unique(levels(taxi$Payment_Type)),
                    ramclass = "factor")
# check result
levels(taxi$Payment_Type)
```

```
## [1] "cash"      "credit"    "dispute"   "no charge"
```

Aggregation with split-apply-combine

First, we will have a look at whether trips paid with credit card tend to involve lower tip amounts than trips paid in cash. In order to do so, we create a table that shows the average amount of tip paid for each payment-type category.

In simple words, this means we first split the dataset into subsets, each of which contains all observations belonging to a distinct payment type. Then, we compute the arithmetic mean of the tip-column of each of these subsets. Finally, we combine all of these results into one table (i.e., the split-apply-combine strategy). When working with `ff`, the `ffdfply()` function in combination with the `doBy` package (Højsgaard and Halekoh, 2023) provides a user-friendly implementation of split-apply-combine types of tasks.

```r
# load packages
library(doBy)


# split-apply-combine procedure on data file chunks
tip_pcategory <- ffdfdply(taxi,
                          split = taxi$Payment_Type,
                          BATCHBYTES = 100000000,
                          FUN = function(x) {
                                summaryBy(Tip_Amt~Payment_Type,
                                        data = x,
                                        FUN = mean,
                                        na.rm = TRUE)})
```

Note how the output describes the procedure step by step. Now we can have a look at the resulting summary statistic in the form of a `data.frame`.

```r
as.data.frame(tip_pcategory)
```

```
##    Payment_Type Tip_Amt.mean
## 1          cash    0.0008162
## 2        credit    2.1619737
## 3       dispute    0.0035075
## 4     no charge    0.0041056
```

The result contradicts our initial hypothesis. However, the comparison is a little flawed. If trips paid by credit card also tend to be longer, the result is not too surprising. We should thus look at the share of tip (or percentage), given the overall amount paid for the trip.

We add an additional variable `percent_tip` and then repeat the aggregation exercise for this variable.

```
# add additional column with the share of tip
taxi$percent_tip <- (taxi$Tip_Amt/taxi$Total_Amt)*100


# recompute the aggregate stats
tip_pcategory <- ffdfdply(taxi,
                          split = taxi$Payment_Type,
                          BATCHBYTES = 100000000,
                          FUN = function(x) {
                              # note the difference here
                              summaryBy(percent_tip~Payment_Type,
                                        data = x,
                                        FUN = mean,
                                        na.rm = TRUE)})
# show result as data frame
as.data.frame(tip_pcategory)


##    Payment_Type percent_tip.mean
## 1          cash         0.005978
## 2        credit        16.004173
## 3       dispute         0.045660
## 4     no charge         0.040433
```

Cross-tabulation of `ff` vectors

Also in relative terms, trips paid by credit card tend to be tipped more. However, are there actually many trips paid by credit card? In order to figure this out, we count the number of trips per payment type by applying the `table.ff` function provided in `ffbase`.

```
table.ff(taxi$Payment_Type)


##
##      cash    credit  dispute no charge
##    781295    215424      536      2745
```

So trips paid in cash are much more frequent than trips paid by credit card. Again using the `table.ff` function, we investigate what factors might be correlated with payment types. First, we have a look at whether payment type is associated with the number of passengers in a trip.

```
# select the subset of observations only containing trips paid by
# credit card or cash
taxi_sub <- subset.ffdf(taxi, Payment_Type=="credit" | Payment_Type == "cash")
taxi_sub$Payment_Type <- ff(taxi_sub$Payment_Type,
                       levels = c("credit", "cash"),
                       ramclass = "factor")


# compute the cross tabulation
crosstab <- table.ff(taxi_sub$Passenger_Count,
                  taxi_sub$Payment_Type
                  )
# add names to the margins
names(dimnames(crosstab)) <- c("Passenger count", "Payment type")
# show result
crosstab
```

```
##                    Payment type
## Passenger count credit     cash
##                0      2      44
##                1 149990 516828
##                2  32891 133468
##                3   7847  36439
##                4   2909  17901
##                5  20688  73027
##                6   1097   3588
```

From the raw numbers it is hard to see whether there are significant differences between the categories cash and credit. We therefore use a visualization technique called a 'mosaic plot' (provided in the vcd package; see Meyer et al. (2023), Meyer et al. (2006), and Zeileis et al. (2007)) to visualize the cross-tabulation.

```
# install.packages(vcd)
# load package for mosaic plot
library(vcd)

# generate a mosaic plot
mosaic(crosstab, shade = TRUE)
```

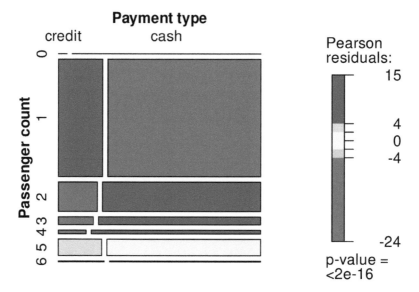

The plot suggests that trips involving more than one passenger tend to be paid by cash rather than by credit card.

10.3   High-speed in-memory data aggregation with `arrow`

For large datasets that (at least in part) fit into RAM, the `arrow` package again provides an attractive alternative to `ff`.

Data import

We use the already familiar `read_csv_arrow()` to import the same first million observations from the taxi trips records.

```
# load packages
library(arrow)
library(dplyr)


# read the CSV file
taxi <- read_csv_arrow("data/tlc_trips.csv",
                       as_data_frame = FALSE)
```

Data preparation and 'split-apply-combine'

We prepare/clean the data as in the `ff`-approach above.

As `arrow` builds on a `dplyr` back-end, basic computations can be easily done through the common `dplyr` syntax. Note, however, that not all of the `dplyr` functions are covered in `arrow` (as of the writing of this book).[3]

```r
# clean the categorical variable; aggregate by group
taxi <-
    taxi %>%
    mutate(Payment_Type = tolower(Payment_Type))


taxi_summary <-
    taxi %>%
    mutate(percent_tip = (Tip_Amt/Total_Amt)*100 ) %>%
    group_by(Payment_Type) %>%
    summarize(avg_percent_tip = mean(percent_tip)) %>%
    collect()
```

Similarly, we can use `data.table`'s `dcast()` for cross-tabulation-like operations.

```r
library(tidyr)

# compute the frequencies; pull result into R
ct <- taxi %>%
    filter(Payment_Type %in% c("credit", "cash")) %>%
    group_by(Passenger_Count, Payment_Type) %>%
    summarize(n=n())%>%
      collect()

# present as cross-tabulation
pivot_wider(data=ct,
            names_from="Passenger_Count",
            values_from = "n")
```

```
## # A tibble: 2 x 11
##    Payment_Type    `1`     `2`     `3`     `4`     `5`     `6`
##    <chr>         <int>   <int>   <int>   <int>   <int>   <int>
## 1 cash         1.42e7  3.57e6  972341  473783  1.89e6   96920
## 2 credit       4.34e6  9.23e5  221648   82800  5.63e5   28853
```

---

[3]If a `dplyr`-like function is not implemented in `arrow`, the `arrow` data object is automatically pulled into R (meaning fully into RAM) and then processed there directly via native `dplyr`. Such a situation might crash your R session due to a lack of RAM.

```
## # i 4 more variables: `0` <int>, `208` <int>,
## #   `129` <int>, `113` <int>
```

10.4 High-speed in-memory data aggregation with `data.table`

For large datasets that still fit into RAM, the `data.table` package (Dowle and Srini-vasan, 2022) provides very fast and elegant functions to compute aggregate statistics.

Data import

We use the already familiar `fread()` to import the same first million observations from the taxi trip records.

```
# load packages
library(data.table)
```

```
# import data into RAM (needs around 200MB)
taxi <- fread("data/tlc_trips.csv",
              nrows = 1000000)
```

Data preparation and `data.table` syntax for 'split-apply-combine'

We prepare/clean the data as in the `ff` approach above.

```
# clean the factor levels
taxi$Payment_Type <- tolower(taxi$Payment_Type)
taxi$Payment_Type <- factor(taxi$Payment_Type,
                            levels = unique(taxi$Payment_Type))
```

Note the simpler syntax of essentially doing the same thing, but all in-memory.

`data.table`-syntax for 'split-apply-combine' operations

With the `[]`-syntax we index/subset the usual `data.frame` objects in R. When work-ing with `data.tables`, much more can be done in the step of 'sub-setting' the frame.[4]

For example, we can directly compute on columns.

---

[4]See https://cran.r-project.org/web/packages/data.table/vignettes/datatable-intro.html for a detailed introduction to the syntax.

```
taxi[, mean(Tip_Amt/Total_Amt)]
```

```
## [1] 0.03452
```

Moreover, in the same step, we can 'split' the rows *by* specific groups and apply the function to each subset.

```
taxi[, .(percent_tip = mean((Tip_Amt/Total_Amt)*100)), by = Payment_Type]
```

```
##      Payment_Type percent_tip
## 1:           cash    0.005978
## 2:         credit   16.004173
## 3:      no charge    0.040433
## 4:        dispute    0.045660
```

Similarly, we can use `data.table`'s `dcast()` for cross-tabulation-like operations.

```
dcast(taxi[Payment_Type %in% c("credit", "cash")],
      Passenger_Count~Payment_Type,
      fun.aggregate = length,
      value.var = "vendor_name")
```

```
##      Passenger_Count   cash credit
## 1:                 0     44      2
## 2:                 1 516828 149990
## 3:                 2 133468  32891
## 4:                 3  36439   7847
## 5:                 4  17901   2909
## 6:                 5  73027  20688
## 7:                 6   3588   1097
```

10.5 Wrapping up

- Similar to the MapReduce idea in the context of distributed systems, the *split-apply-combine* approach is key in many Big Data aggregation procedures on normal machines (laptop/desktop computers). The idea is to split the overall data into subsets based on a categorical variable, apply a function (e.g., mean) on each subset, and then combine the results into one object. Thus, the approach allows for parallelization and working on separate data chunks.

- As computing descriptive statistics on various subsets of a large dataset can be very memory-intensive, it is recommended to use out-of-memory strategies, lazy evaluation, or a classical SQL-database approach for this.
- There are several options available such as `ffdply`, running on chunked datasets; and `arrow` with `group_by()`.

11

(Big) Data Visualization

Visualizing certain characteristics and patterns in large datasets is primarily challenging for two reasons. First, depending on the type of plot, plotting raw data consisting of many observations can take a long time (and lead to large figure files). Second, patterns might be harder to recognize due to the sheer amount of data displayed in a plot. Both of these challenges are particularly related to the visualization of raw data for explorative or descriptive purposes. Visualizations of already-computed aggregations or estimates is typically very similar whether working with large or small datasets.

The following sections thus particularly highlight the issue of generating plots based on the raw data, including many observations, and with the aim of exploring the data in order to discover patterns in the data that then can be further investigated in more sophisticated statistical analyses. We will do so in three steps. First, we will look into a few important conceptual aspects where generating plots with a large number of observations becomes difficult, and then we will look at potentially helpful tools to address these difficulties. Based on these insights, the next section presents a data exploration tutorial based on the already familiar TLC taxi trips dataset, looking into different approaches to visualize relations between variables. Finally, the last section of this chapter covers an area of data visualization that has become more and more relevant in applied economic research with the availability of highly detailed observational data on economic and social activities (due to the digitization of many aspects of modern life): the plotting of geo-spatial information on economic activity.

All illustrations of concepts and visualization examples in this chapter build on the Grammar of Graphics (Wilkinson et al., 2005) concept implemented in the `ggplot2` package (Wickham, 2016). The choice of this plotting package/framework is motivated by the large variety of plot-types covered in `ggplot2` (ranging from simple scatterplots to hexbin-plots and geographic maps), as well as the flexibility to build and modify plots step by step (an aspect that is particularly interesting when exploring large datasets visually).

11.1   Challenges of Big Data visualization

Generating a plot in an interactive R session means generating a new object in the
R environment (RAM), which can (in the case of large datasets) take up a consider-
able amount of memory. Moreover, depending on how the plot function is called,
RStudio will directly render the plot in the Plots tab (which again needs memory
and processing). Consider the following simple example, in which we plot two vectors of random numbers against each other.[1]

```
# load package
library(ggplot2) # for plotting
library(pryr) # for profiling
library(bench) # for profiling
library(fs) # for profiling


# random numbers generation
x <- rnorm(10^6, mean=5)
y <- 1 + 1.4*x + rnorm(10^6)
plotdata <- data.frame(x=x, y=y)
object_size(plotdata)


## 16.00 MB


# generate scatter plot
splot <-
    ggplot(plotdata, aes(x=x, y=y))+
    geom_point()
object_size(splot)


## 16.84 MB
```

The plot object, not surprisingly, takes up an additional slice of RAM of the size
of the original dataset, plus some overhead. Now when we instruct ggplot to gen-
erate/plot the visualization on canvas, even more memory is needed. Moreover,
rather a lot of data processing is needed to place one million points on the canvas
(also, note that one million observations would not be considered a lot in the context of this book...).

---

[1]We will use bench and the fs package (Hester et al., 2023) for profiling.

```
mem_used()
```

```
## 2.26 GB
```

```
system.time(print(splot))
```

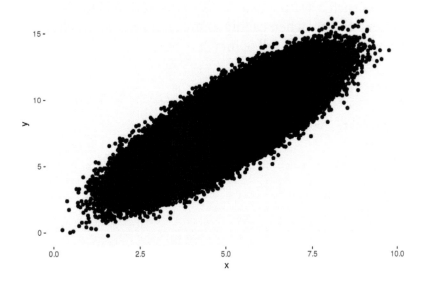

```
##    user  system elapsed
##   2.583   0.710   3.306
```

```
mem_used()
```

```
## 2.36 GB
```

First, to generate this one plot, an average modern laptop needs about 13.6 seconds. This would not be very comfortable in an interactive session to explore the data visually. Second, and even more striking, before the plot was generated, mem_used() indicated the total amount of memory (in MBs) used by R was around 160MB, while right after plotting to the canvas, R had used around 270MB. Note that this is larger than the dataset and the ggplot-object by an order of magnitude. Creating the same plot based on 100 million observations would likely crash or freeze your R session. Finally, when we output the plot to a file (for example, a pdf), the generated vector-based graphic file is also rather large.

```
ggsave("splot.pdf", device="pdf", width = 5, height = 5)
file_size("splot.pdf")
```

```
## 54.8M
```

Hence generating plots visualizing large amounts of raw data tends to use up a lot of computing time, memory, and (ultimately) storage space for the generated plot file. There are a couple of solutions to address these performance issues.

Avoid fancy symbols (costly rendering) It turns out that one aspect of the problem is the particular symbols/characters used in ggplot (and other plot functions in R) for the points in such a scatter-plot. Thus, one solution is to override the default set of characters directly when calling `ggplot()`. A reasonable choice of character for this purpose is simply the full point (.).

```
# generate scatter plot
splot2 <-
     ggplot(plotdata, aes(x=x, y=y))+
     geom_point(pch=".")


mem_used()


## 2.26 GB


system.time(print(splot2))
```

```
##     user  system elapsed
##    1.856   0.260   2.121
```

```
mem_used()
```

```
## 2.37 GB
```

The increase in memory due to the plot call is comparatively smaller, and plotting is substantially faster.

Use rasterization (bitmap graphics) instead of vector graphics By default, most data visualization libraries, including ggplot2, are implemented to generate vector-based graphics. Conceptually, this makes a lot of sense for any type of plot when the number of observations plotted is small or moderate. In simple terms, vector-based graphics define lines and shapes as vectors in a coordinate system. In the case of a scatter-plot, the x and y coordinates of every point need to be recorded. In contrast, bitmap files contain image information in the form of a matrix (or several matrices if colors are involved), whereby each cell of the matrix represents a pixel and contains information about the pixel's color. While a vector-based representation of plot of few observations is likely more memory-efficient than a high-resolution bitmap representation of the same plot, it might well be the other way around when we are plotting millions of observations.

Thus, an alternative solution to save time and memory is to directly use a bitmap format instead of a vector-based format. This could be done by plotting directly to a bitmap-format file and then opening the file to look at the plot. However, this is somewhat clumsy as part of a data visualization workflow to explore the data. Luckily there is a ready-made solution by Kratochvíl et al. (2020) that builds on the idea of rasterizing scatter-plots, but that then displays the bitmap image directly in R. The approach is implemented in the scattermore package (Kratochvil, 2022) and can straightforwardly be used in combination with ggplot.

```
# install.packages("scattermore")
library(scattermore)
# generate scatter plot
splot3 <-
    ggplot()+
    geom_scattermore(aes(x=x, y=y), data=plotdata)

# show plot in interactive session
system.time(print(splot3))
```

```
##     user  system elapsed
##    0.678   0.020   0.705
```

```
# plot to file
ggsave("splot3.pdf",  device="pdf", width = 5, height = 5)
file_size("splot3.pdf")
```

```
## 13.2K
```

This approach is faster by an order of magnitude, and the resulting pdf takes up only a fraction of the storage space needed for splot.pdf, which is based on the classical geom_points() and a vector-based image.

Use aggregates instead of raw data Depending on what pattern/aspect of the data you want to inspect visually, you might not actually need to plot all observations directly but rather the result of aggregating the observations first. There are several options to do this, but in the context of scatter plots based on many observations, a two-dimensional bin plot can be a good starting point. The idea behind this approach is to divide the canvas into grid-cells (typically in the form of rectangles or hexagons), compute for each grid cell the number of observations/points that would fall into it (in a scatter plot), and then indicate the number of observations per grid cell via the cell's shading. Such a 2D bin plot of the same data as above can be generated via geom_hex():

```
# generate scatter plot
splot4 <-
```

```
ggplot(plotdata, aes(x=x, y=y))+
geom_hex()
```

```
mem_used()
```

```
## 2.26 GB
```

```
system.time(print(splot4))
```

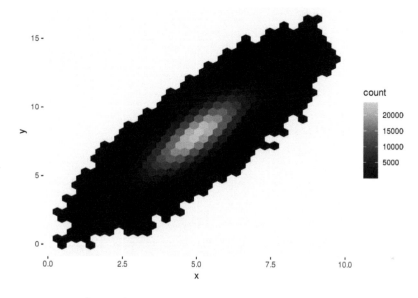

```
##     user  system elapsed
##    0.471   0.001   0.494
```

```
mem_used()
```

```
## 2.27 GB
```

Obviously, this approach is much faster and uses up much less memory than the `geom_point()` approach. Moreover, note that this approach to visualizing a potential relation between two variables based on many observations might even have another advantage over the approaches taken above. In all of the scatter plots, it was not visible whether the point cloud contains areas with substantially more observations (more density). There were simply too many points plotted over each other to recognize much more than the contour of the overall point cloud. With the 2D bin plot implemented with `geom_hex()`, we recognize immediately that there are many more observations located in the center of the cloud than further away from the center.

11.2   Data exploration with ggplot2

In this tutorial we will work with the TLC data used in the data aggregation session. The raw data consists of several monthly CSV files and can be downloaded via the TLC's website[2]. Again, we work only with the first million observations.

In order to better understand the large dataset at hand (particularly regarding the determinants of tips paid), we use ggplot2 to visualize some key aspects of the data.

First, let's look at the raw relationship between fare paid and the tip paid. We set up the canvas with ggplot.

```
# load packages
library(ggplot2)

# set up the canvas
taxiplot <- ggplot(taxi, aes(y=Tip_Amt, x= Fare_Amt))
taxiplot
```

Now we visualize the co-distribution of the two variables with a simple scatter-plot. to speed things up, we use geom_scattermore() but increase the point size.[3]

---

[2]https://www1.nyc.gov/site/tlc/about/tlc-trip-record-data.page

[3]Note how the points look less nice than what geom_point() would produce. This is the disadvantage of using the bitmap approach rather than the vector-based approach.

```
# simple x/y plot
taxiplot + geom_scattermore(pointsize = 3)
```

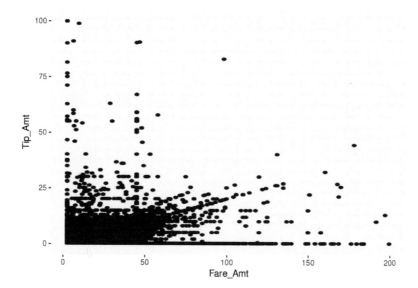

Note that this took quite a while, as R had to literally plot one million dots on the canvas. Moreover, many dots fall within the same area, making it impossible to recognize how much mass there actually is. This is typical for visualization exercises with large datasets. One way to improve this is by making the dots more transparent by setting the `alpha` parameter.

```
# simple x/y plot
taxiplot + geom_scattermore(pointsize = 3, alpha=0.2)
```

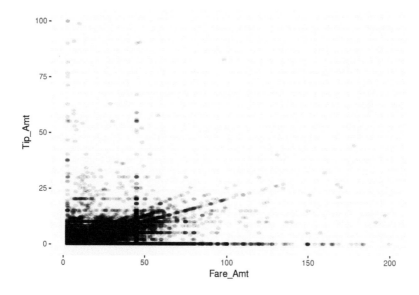

Alternatively, we can compute two-dimensional bins. Here, we use `geom_bin2d()` (an alternative to `geom_hex` used above) in which the canvas is split into rectangles and the number of observations falling into each respective rectangle is computed. The visualization is then based on plotting the rectangles with counts greater than 0, and the shading of the rectangles indicates the count values.

```
# two-dimensional bins
taxiplot + geom_bin2d()
```

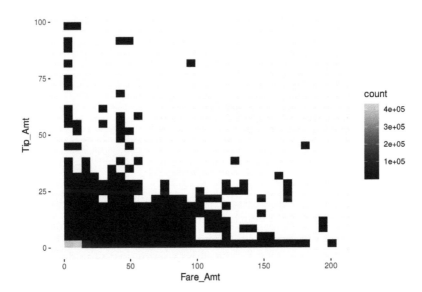

A large proportion of the tip/fare observations seem to be in the very lower-left corner of the pane, while most other trips seem to be evenly distributed. However, we fail to see smaller differences in this visualization. In order to reduce the dominance of the 2D bins with very high counts, we display the natural logarithm of counts and display the bins as points.

```
# two-dimensional bins
taxiplot +
    stat_bin_2d(geom="point",
                mapping= aes(size = log(after_stat(count)))) +
    guides(fill = "none")
```

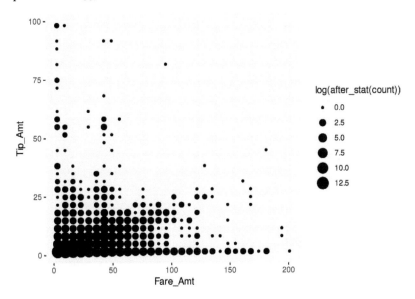

We note that there are many cases with very low fare amounts, many cases with no or hardly any tip, and quite a lot of cases with very high tip amounts (in relation to the rather low fare amount). In the following, we dissect this picture by having a closer look at 'typical' tip amounts and whether they differ by type of payment.

```r
# compute frequency of per tip amount and payment method
taxi[, n_same_tip:= .N, by= c("Tip_Amt", "Payment_Type")]
frequencies <- unique(taxi[Payment_Type %in% c("credit", "cash"),
                       c("n_same_tip",
                         "Tip_Amt",
                         "Payment_Type")][order(n_same_tip,
                                          decreasing = TRUE)])


# plot top 20 frequent tip amounts
fare <- ggplot(data = frequencies[1:20], aes(x = factor(Tip_Amt),
                                       y = n_same_tip))
fare + geom_bar(stat = "identity")
```

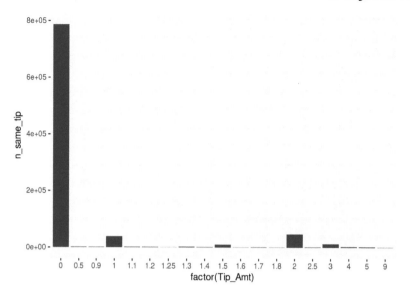

Indeed, paying no tip at all is quite frequent, overall.[4] The bar plot also indicates that there seem to be some 'focal points' in the amount of tip paid. Clearly, paying one USD or two USD is more common than paying fractions. However, fractions of dollars might be more likely if tips are paid in cash and customers simply add some loose change to the fare amount paid.

```
fare + geom_bar(stat = "identity") +
    facet_wrap("Payment_Type")
```

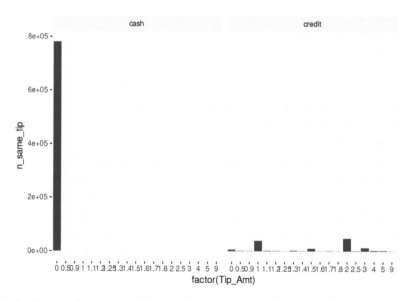

Clearly, it looks as if trips paid in cash tend not to be tipped (at least in this sub-sample).

---

[4]Or, could there be another explanation for this pattern in the data?

Let's try to tease this information out of the initial points plot. Trips paid in cash are often not tipped; we thus should indicate the payment method. Moreover, tips paid in full dollar amounts might indicate a habit.

```
# indicate natural numbers
taxi[, dollar_paid := ifelse(Tip_Amt == round(Tip_Amt,0), "Full", "Fraction"),]
```

```
# extended x/y plot
taxiplot +
    geom_scattermore(pointsize = 3, alpha=0.2, aes(color=Payment_Type)) +
    facet_wrap("dollar_paid") +
    theme(legend.position="bottom")
```

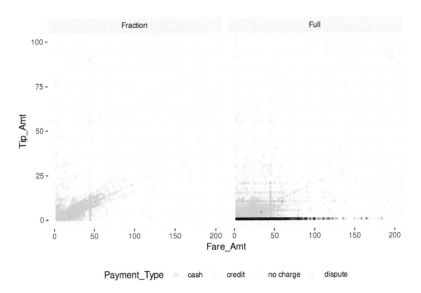

Now the picture is getting clearer. Paying a tip seems to follow certain rules of thumb. Certain fixed amounts tend to be paid independent of the fare amount (visible in the straight lines of dots on the right-hand panel). At the same time, the pattern in the left panel indicates another habit: computing the amount of the tip as a linear function of the total fare amount ('pay 10% tip'). A third habit might be to determine the amount of tip by 'rounding up' the total amount paid. In the following, we try to tease the latter out, only focusing on credit card payments.

```
taxi[, rounded_up := ifelse(Fare_Amt + Tip_Amt == round(Fare_Amt + Tip_Amt, 0),
                            "Rounded up",
                            "Not rounded")]
# extended x/y plot
taxiplot +
```

```
geom_scattermore(data= taxi[Payment_Type == "credit"],
                 pointsize = 3, alpha=0.2, aes(color=rounded_up)) +
facet_wrap("dollar_paid") +
theme(legend.position="bottom")
```

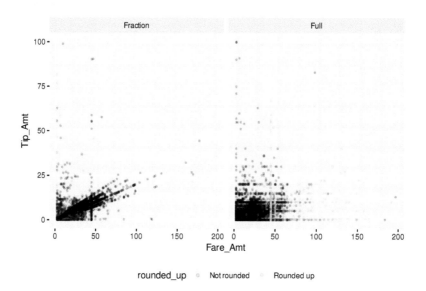

Now we can start modeling. A reasonable first shot is to model the tip amount as a linear function of the fare amount, conditional on no-zero tip amounts paid as fractions of a dollar.

```
modelplot <- ggplot(data= taxi[Payment_Type == "credit" &
                               dollar_paid == "Fraction" &
                               0 < Tip_Amt],
                    aes(x = Fare_Amt, y = Tip_Amt))
modelplot +
    geom_scattermore(pointsize = 3, alpha=0.2, color="darkgreen") +
    geom_smooth(method = "lm", colour = "black")   +
    theme(legend.position="bottom")
```

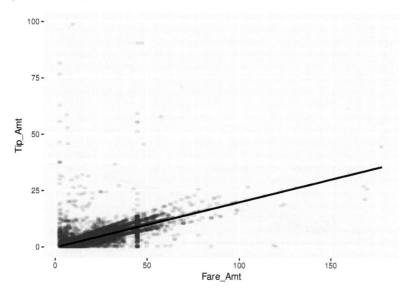

Finally, we prepare the plot for reporting. `ggplot2` provides several predefined 'themes' for plots that define all kinds of aspects of a plot (background color, line colors, font size, etc.). The easiest way to tweak the design of your final plot in a certain direction is to just add such a pre-defined theme at the end of your plot. Some of the pre-defined themes allow you to change a few aspects, such as the font type and the base size of all the texts in the plot (labels, tick numbers, etc.). Here, we use the `theme_bw()`, increase the font size, and switch to a serif-type font. `theme_bw()` is one of the complete themes that ships with the basic `ggplot2` installation.[5] Many more themes can be found in additional R packages (see, for example, the `ggthemes` package[6]).

```
modelplot <- ggplot(data= taxi[Payment_Type == "credit"
                          & dollar_paid == "Fraction"
                          & 0 < Tip_Amt],
                 aes(x = Fare_Amt, y = Tip_Amt))
modelplot +
    geom_scattermore(pointsize = 3, alpha=0.2, color="darkgreen") +
    geom_smooth(method = "lm", colour = "black") +
    ylab("Amount of tip paid (in USD)") +
    xlab("Amount of fare paid (in USD)") +
    theme_bw(base_size = 18, base_family = "serif")
```

[5]See the ggplot2 documentation (https://ggplot2.tidyverse.org/reference/ggtheme.html) for a list of all pre-defined themes shipped with the basic installation.
[6]https://cran.r-project.org/web/packages/ggthemes/index.html

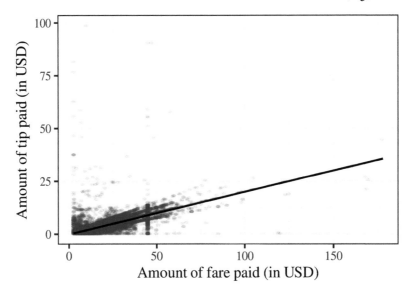

Aside: modify and create themes

Simple modifications of themes

Apart from using pre-defined themes as illustrated above, we can use the `theme()` function to further modify the design of a plot. For example, we can print the axis labels ('axis titles') in bold.

```r
modelplot <- ggplot(data= taxi[Payment_Type == "credit"
                               & dollar_paid == "Fraction"
                               & 0 < Tip_Amt],
                aes(x = Fare_Amt, y = Tip_Amt))
modelplot +
    geom_scattermore(pointsize = 3, alpha=0.2, color="darkgreen") +
    geom_smooth(method = "lm", colour = "black") +
    ylab("Amount of tip paid (in USD)") +
    xlab("Amount of fare paid (in USD)") +
    theme_bw(base_size = 18, base_family = "serif") +
    theme(axis.title = element_text(face="bold"))
```

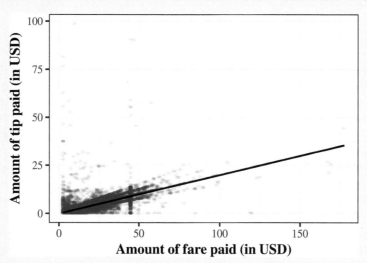

There is a large list of plot design aspects that can be modified in this way (see ?theme() for details).

Create your own themes

Extensive design modifications via theme() can involve many lines of code, making your plot code harder to read/understand. In practice, you might want to define your specific theme once and then apply this theme to all of your plots. In order to do so it makes sense to choose one of the existing themes as a basis and then modify its design aspects until you have the design you are looking for. Following the design choices in the examples above, we can create our own theme_my_serif() as follows.

```r
# 'define' a new theme
theme_my_serif <-
  theme_bw(base_size = 18, base_family = "serif") +
  theme(axis.title = element_text(face="bold"))


# apply it
modelplot +
    geom_scattermore(pointsize = 3, alpha=0.2, color="darkgreen") +
    geom_smooth(method = "lm", colour = "black") +
    ylab("Amount of tip paid (in USD)") +
    xlab("Amount of fare paid (in USD)") +
  theme_my_serif
```

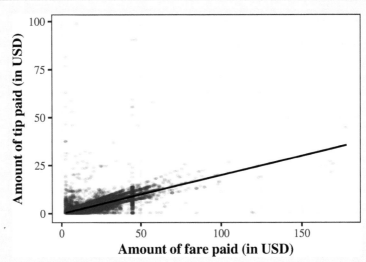

This practical approach does not require you to define every aspect of a theme. If you indeed want to completely define every aspect of a theme, you can set complete=TRUE when calling the theme function.

```
# 'define' a new theme
my_serif_theme <-
   theme_bw(base_size = 18, base_family = "serif") +
   theme(axis.title = element_text(face="bold"), complete = TRUE)

# apply it
modelplot +
    geom_scattermore(pointsize = 3, alpha=0.2, color="darkgreen") +
    geom_smooth(method = "lm", colour = "black") +
    ylab("Amount of tip paid (in USD)") +
    xlab("Amount of fare paid (in USD)") +
   theme_my_serif
```

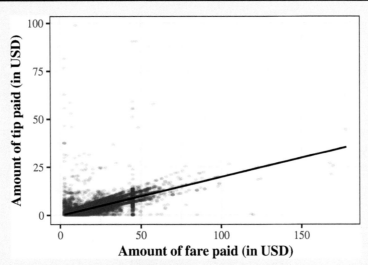

Note that since we have only defined one aspect (bold axis titles), the rest of the elements follow the default theme.

Implementing actual themes as functions

Importantly, the approach outlined above does not technically really create a new theme like `theme_bw()`, as these pre-defined themes are implemented as functions. Note that we add the new theme to the plot simply with + `theme_my_serif` (no parentheses). In practice this is the simplest approach, and it provides all the functionality you need in order to apply your own 'theme' to each of your plots. If you want to implement a theme as a function, the following blueprint can get you started.

```r
# define own theme
theme_my_serif <-
  function(base_size = 15,
           base_family = "",
           base_line_size = base_size/170,
           base_rect_size = base_size/170){
    # use theme_bw() as a basis but replace some design elements
    theme_bw(base_size = base_size,
             base_family = base_family,
             base_line_size = base_size/170,
             base_rect_size = base_size/170) %+replace%
      theme(
        axis.title = element_text(face="bold")
      )
  }


# apply the theme
modelplot +
    geom_scattermore(pointsize = 3, alpha=0.2, color="darkgreen") +
    geom_smooth(method = "lm", colour = "black") +
    ylab("Amount of tip paid (in USD)") +
    xlab("Amount of fare paid (in USD)") +
  theme_my_serif(base_size = 18, base_family="serif")
```

11.3   Visualizing time and space

The previous visualization exercises were focused on visually exploring patterns in the tipping behavior of people taking a NYC yellow cab ride. Based on the same dataset, we will explore the time and spatial dimensions of the TLC Yellow Cab data. That is, we explore where trips tend to start and end, depending on the time of the day.

11.3.1   Preparations

For the visualization of spatial data, we first load additional packages that give R some GIS features.

```
# load GIS packages
library(rgdal)
library(rgeos)
```

Moreover, we download and import a so-called 'shape file'[7] (a geospatial data format) of New York City. This will be the basis for our visualization of the spatial dimension of taxi trips. The file is downloaded from New York's Department of City Planning[8] and indicates the city's community district borders.[9]

```
# download the zipped shapefile to a temporary file; unzip
BASE_URL <-
"https://www1.nyc.gov/assets/planning/download/zip/data-maps/open-data/"
FILE <- "nycd_19a.zip"
URL <- paste0(BASE_URL, FILE)
tmp_file <- tempfile()
download.file(URL, tmp_file)
file_path <- unzip(tmp_file, exdir= "data")
# delete the temporary file
unlink(tmp_file)
```

Now we can import the shape file and have a look at how the GIS data is structured.

```
# read GIS data
nyc_map <- readOGR(file_path[1], verbose = FALSE)
# have a look at the GIS data
summary(nyc_map)
```

```
## Object of class SpatialPolygonsDataFrame
## Coordinates:
##       min      max
## x 913175 1067383
## y 120122  272844
## Is projected: TRUE
## proj4string :
## [+proj=lcc +lat_0=40.1666666666667 +lon_0=-74
## +lat_1=41.0333333333333 +lat_2=40.6666666666667
```

---

[7]https://en.wikipedia.org/wiki/Shapefile

[8]https://www1.nyc.gov/site/planning/index.page

[9]Similar files are provided online by most city authorities in developed countries. See, for example, GIS Data for the City and Canton of Zurich: https://maps.zh.ch/.

```
## +x_0=300000 +y_0=0 +datum=NAD83 +units=us-ft
## +no_defs]
## Data attributes:
##       BoroCD         Shape_Leng          Shape_Area
##   Min.    :101    Min.    : 23963    Min.    :2.43e+07
##   1st Qu.:206    1st Qu.: 36611    1st Qu.:4.84e+07
##   Median :308    Median : 52246    Median :8.27e+07
##   Mean    :297    Mean    : 74890    Mean    :1.19e+08
##   3rd Qu.:406    3rd Qu.: 85711    3rd Qu.:1.37e+08
##   Max.    :595    Max.    :270660    Max.    :5.99e+08
```

Note that the coordinates are not in the usual longitude and latitude units. The original map uses a different projection than the TLC data of taxi trip records. Before plotting, we thus have to change the projection to be in line with the TLC data.

```
# transform the projection
p <- CRS("+proj=longlat +datum=WGS84 +no_defs +ellps=WGS84 +towgs84=0,0,0")
nyc_map <- spTransform(nyc_map, p)
# check result
summary(nyc_map)
```

```
## Object of class SpatialPolygonsDataFrame
## Coordinates:
##       min     max
## x -74.26 -73.70
## y   40.50   40.92
## Is projected: FALSE
## proj4string : [+proj=longlat +datum=WGS84 +no_defs]
## Data attributes:
##       BoroCD         Shape_Leng          Shape_Area
##   Min.    :101    Min.    : 23963    Min.    :2.43e+07
##   1st Qu.:206    1st Qu.: 36611    1st Qu.:4.84e+07
##   Median :308    Median : 52246    Median :8.27e+07
##   Mean    :297    Mean    : 74890    Mean    :1.19e+08
##   3rd Qu.:406    3rd Qu.: 85711    3rd Qu.:1.37e+08
##   Max.    :595    Max.    :270660    Max.    :5.99e+08
```

One last preparatory step is to convert the map data to a `data.frame` for plotting with `ggplot`.

```
nyc_map <- fortify(nyc_map)
```

11.3.2 Pick-up and drop-off locations

Since trips might actually start or end outside of NYC, we first restrict the sample of trips to those within the boundary box of the map. For the sake of the exercise, we only select a random sample of 50000 trips from the remaining trip records.

```
# taxi trips plot data
taxi_trips <- taxi[Start_Lon <= max(nyc_map$long) &
                    Start_Lon >= min(nyc_map$long) &
                    End_Lon <= max(nyc_map$long) &
                    End_Lon >= min(nyc_map$long) &
                    Start_Lat <= max(nyc_map$lat) &
                    Start_Lat >= min(nyc_map$lat) &
                    End_Lat <= max(nyc_map$lat) &
                    End_Lat >= min(nyc_map$lat)
                    ]
taxi_trips <- taxi_trips[base::sample(1:nrow(taxi_trips), 50000)]
```

In order to visualize how the cab traffic is changing over the course of the day, we add an additional variable called `start_time` in which we store the time (hour) of the day a trip started.

```
taxi_trips$start_time <- lubridate::hour(taxi_trips$Trip_Pickup_DateTime)
```

Particularly, we want to look at differences between morning, afternoon, and evening/night.

```
# define new variable for facets
taxi_trips$time_of_day <- "Morning"
taxi_trips[start_time > 12 & start_time < 17]$time_of_day <- "Afternoon"
taxi_trips[start_time %in% c(17:24, 0:5)]$time_of_day <- "Evening/Night"
taxi_trips$time_of_day <-
  factor(taxi_trips$time_of_day,
        levels = c("Morning", "Afternoon", "Evening/Night"))
```

We create the plot by first setting up the canvas with our taxi trip data. Then, we add the map as a first layer.

```
# set up the canvas
locations <- ggplot(taxi_trips, aes(x=long, y=lat))
# add the map geometry
```

```
locations <- locations + geom_map(data = nyc_map,
                                  map = nyc_map,
                                  aes(map_id = id))
locations
```

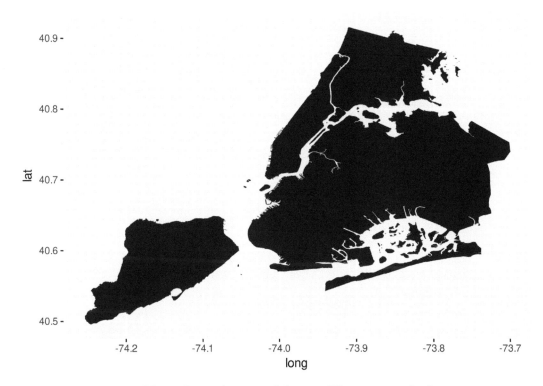

Now we can start adding the pick-up and drop-off locations of cab trips.

```
# add pick-up locations to plot
locations +
    geom_scattermore(aes(x=Start_Lon, y=Start_Lat),
            color="orange",
            pointsize = 1,
            alpha = 0.2)
```

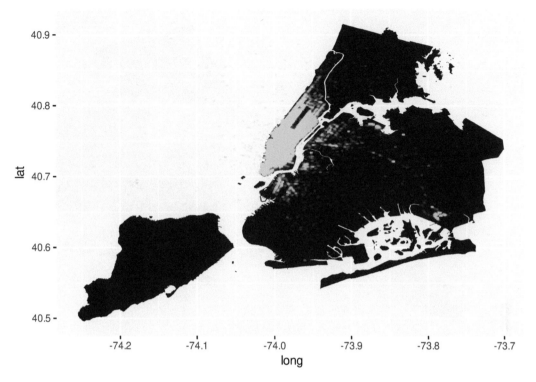

As is to be expected, most of the trips start in Manhattan. Now let's look at where trips end.

```
# add drop-off locations to plot
locations +
    geom_scattermore(aes(x=End_Lon, y=End_Lat),
            color="steelblue",
            pointsize = 1,
            alpha = 0.2) +
    geom_scattermore(aes(x=Start_Lon, y=Start_Lat),
            color="orange",
            pointsize = 1,
            alpha = 0.2)
```

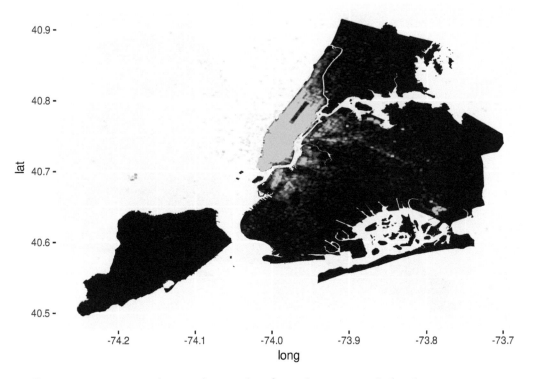

In fact, more trips tend to end outside of Manhattan. And the destinations seem to be broader spread across the city then the pick-up locations. Most destinations are still in Manhattan, though.

Now let's have a look at how this picture changes depending on the time of the day.

```
# pick-up locations
locations +
    geom_scattermore(aes(x=Start_Lon, y=Start_Lat),
            color="orange",
            pointsize =1,
            alpha = 0.2) +
    facet_wrap(vars(time_of_day))
```

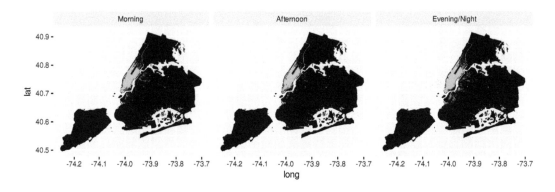

```
# drop-off locations
locations +
    geom_scattermore(aes(x=End_Lon, y=End_Lat),
                color="steelblue",
                pointsize = 1,
                alpha = 0.2) +
    facet_wrap(vars(time_of_day))
```

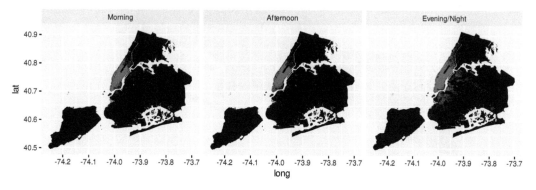

Alternatively, we can plot the hours on a continuous scale.

```
# drop-off locations
locations +
    geom_scattermore(aes(x=End_Lon, y=End_Lat, color = start_time),
                pointsize = 1,
                alpha = 0.2) +
    scale_colour_gradient2( low = "red", mid = "yellow", high = "red",
                        midpoint = 12)
```

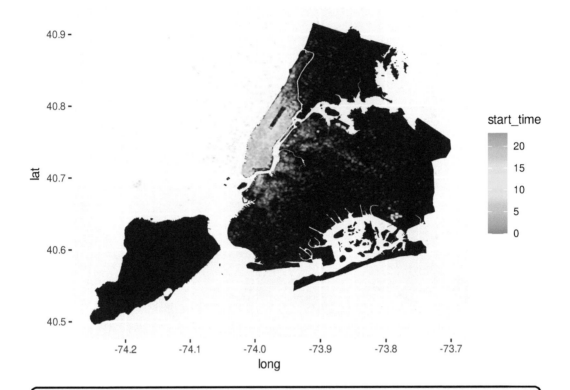

Aside: change color schemes

In the example above we use scale_colour_gradient2() to modify the color gradient used to visualize the start time of taxi trips. By default, ggplot would plot the following (default gradient color setting):

```
# drop-off locations
locations +
      geom_scattermore(aes(x=End_Lon, y=End_Lat, color = start_time ),
                 pointsize = 1,
                 alpha = 0.2)
```

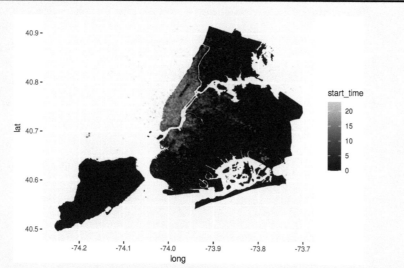

ggplot2 offers various functions to modify the color scales used in a plot. In the case of the example above, we visualize values of a continuous variable. Hence we use a gradient color scale. In the case of categorical variables, we need to modify the default discrete color scale.

Recall the plot illustrating tipping behavior, where we highlight in which observations the client paid with credit card, cash, etc.

```r
# indicate natural numbers
taxi[, dollar_paid := ifelse(Tip_Amt == round(Tip_Amt,0),
                             "Full",
                             "Fraction"),]
# extended x/y plot
taxiplot +
    geom_scattermore(alpha=0.2,
                     pointsize=3,
                     aes(color=Payment_Type)) +
    facet_wrap("dollar_paid") +
    theme(legend.position="bottom")
```

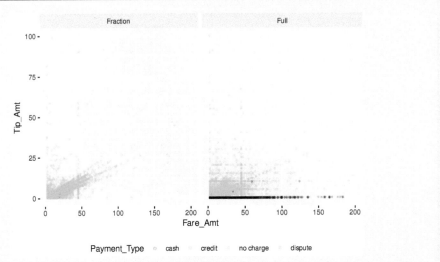

Since we do not further specify the discrete color scheme to be used, ggplot simply uses its default color scheme for this plot. We can change this as follows.

```
# indicate natural numbers
taxi[, dollar_paid := ifelse(Tip_Amt == round(Tip_Amt,0),
                             "Full",
                             "Fraction"),]
# extended x/y plot
taxiplot +
    geom_scattermore(alpha=0.2, pointsize = 3,
                     aes(color=Payment_Type)) +
    facet_wrap("dollar_paid") +
    scale_color_discrete(type = c("red",
                                  "steelblue",
                                  "orange",
                                  "purple")) +
    theme(legend.position="bottom")
```

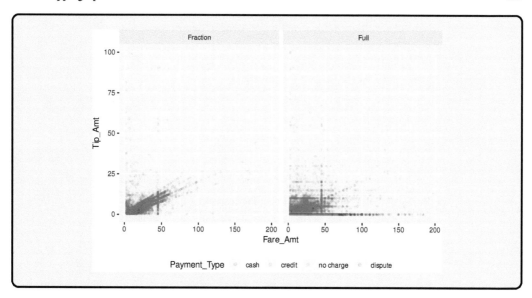

11.4 Wrapping up

- *ggplot* offers a unified approach to generating a variety of plots common in the Big Data context: heatmaps, GIS-like maps, density plots, 2D-bin plots, etc.
- Building on the concept of the *Grammar of Graphics* (Wilkinson et al., 2005), ggplot2 follows the paradigm of creating plots layer-by-layer, which offers great flexibility regarding the visualization of complex (big) data.
- Standard plotting facilities in R (including in ggplot) are based on the concept of vector images (where each dot, line, and area is defined as in a coordinate system). While vector images have the advantage of flexible scaling (no reliance on a specific resolution), when plotting many observations, the computational load to generate and store/hold such graphics in memory can be substantial.
- Plotting of large amounts of data can be made more efficient by relying on less complex shapes (e.g., for dots in a scatter-plot) or through *rasterization* and conversion of the plot into a *bitmap-image (a raster-based image)*. In contrast to vector images, raster images are created with a specific resolution that defines the size of a matrix of pixels that constitutes the image. If plotting a scatter-plot based on many observations, this data structure is much more memory-efficient than defining each dot in a vector image.
- Specific types of plots, such as hex-bin plots and other 2D-bin plots, facilitate plotting large amounts of data independent of the type of image (vector or raster). Moreover, they can be useful to show/highlight specific patterns in large amounts of data that could not be seen in standard scatter plots.

Part IV

Application: Topics in Big Data Econometrics

Introduction

Far from being a comprehensive overview of Big Data Analytics applications in applied econometrics, the goal of this part of the book is to connect the conceptual and practical material covered thus far with analytics settings and point you to potentially interesting approaches and tools that may be useful in your specific field of applying Big Data Analytics. While all of the examples, case studies, and tutorials presented in this part are in the context of economic research or business analytics, the approaches and tools discussed are often easily transferable to other domains of applying Big Data Analytics.

First, Chapter 12 presents a few brief case studies pointing to a small exemplary selection of common bottlenecks in everyday data analytics tasks, such as the estimation of fixed effects models. The purpose of this chapter is to review some of the key concepts discussed in the previous two parts, whereby each of the case studies refers to some of the previously discussed perspectives on/approaches to Big Data: realizing why an analytics task is a burden for the available resources and considering an alternative statistical procedure, writing efficient R code for simple computations, and scaling up the computing resources. You can easily skip this chapter if you are already well familiar with the topics covered in the previous parts. Chapters 13–15 cover specific topic domains in the realm of applied Big Data Analytics that are common in modern econometrics: training machine learning models in predictive econometrics using GPUs (Chapter 13), estimating linear and generalized linear models (e.g., classification models) with large scale datasets (Chapter 14), and performing large-scale text analysis (Chapter 15).

12

Bottlenecks in Everyday Data Analytics Tasks

This chapter presents three examples of how the lessons from the previous chapters could be applied in everyday data analytics tasks. The first section focuses on the statistics perspective: compute something in a different way (with a different algorithm) but end up with essentially the same result. It is also an illustration of how diverse the already implemented solutions for working with large data in applied econometrics in the R-universe are, and how it makes sense to first look into a more efficient algorithm/statistical procedure than directly use specialized packages such as `bigmemory` or even scale up in the cloud. The second section is a reminder (in an extremely simple setting) of how we can use R more efficiently when taking basic R characteristics into consideration. It is an example and detailed illustration of how adapting a few simple coding habits with basic R can substantially improve the efficiency of your code for larger workloads. Finally, the third section in this chapter re-visits the topics of scaling up both locally and in the cloud.

12.1   Case study: Efficient fixed effects estimation

In this case study we look into a very common computational problem in applied econometrics: estimation of a fixed effects model with various fixed-effects units (i.e., many intercepts). The aim of this case study is to give an illustration of how a specific statistical procedure can help us reduce the computational burden substantially (here, by reducing the number of columns in the model matrix and therefore the burden of computing the inverse of a huge model matrix). The context of this tutorial builds on a study called "Friends in High Places"[1] by Cohen and Malloy (2014). Cohen and Malloy show that US Senators who are alumni of the same university/college tend to help each other out in votes on industrial policies if the corresponding policy is highly relevant for the state of one senator but not relevant for the state of the other senator. The data is provided along with the published article and can be accessed here: http://doi.org/10.3886/E114873V1. The data (and code) is provided in STATA format. We can import the main dataset with the `foreign`

---

[1]https://www.aeaweb.org/articles?id=10.1257/pol.6.3.63

package (R Core Team, 2022). For data handling we load the `data.table` package and for hypotheses tests we load the `lmtest` package (Zeileis and Hothorn, 2002).

```
# SET UP -------------------
# load packages
library(foreign)
library(data.table)
library(lmtest)
# fix vars
DATA_PATH <- "data/data_for_tables.dta"

# import data
cm <- as.data.table(read.dta(DATA_PATH))
# keep only clean obs
cm <- cm[!(is.na(yes)
          |is.na(pctsumyessameparty)
          |is.na(pctsumyessameschool)
          |is.na(pctsumyessamestate))]
```

As part of this case study, we will replicate parts of Table 3 of the main article (p. 73). Specifically, we will estimate specifications (1) and (2). In both specifications, the dependent variable is an indicator `yes` that is equal to 1 if the corresponding senator voted Yes on the given bill and 0 otherwise. The main explanatory variables of interest are `pctsumyessameschool` (the percentage of senators from the same school as the corresponding senator who voted Yes on the given bill), `pctsumyessamestate` (the percentage of senators from the same state as the corresponding senator who voted Yes on the given bill), and `pctsumyessameparty` (the percentage of senators from the same party as the corresponding senator who voted Yes on the given bill). Specification 1 accounts for congress (time) fixed effects and senator (individual) fixed effects, and specification 2 accounts for congress-session-vote fixed effects and senator fixed effects.

First, let us look at a very simple example to highlight where the computational burden in the estimation of such specifications is coming from. In terms of the regression model 1, the fixed effect specification means that we introduce an indicator variable (an intercept) for $N - 1$ senators and $M - 1$ congresses. That is, the simple model matrix (X) without accounting for fixed effects has dimensions 425653×4.

```
# pooled model (no FE)
model0 <-    yes ~
```

```
    pctsumyessameschool +
    pctsumyessamestate +
    pctsumyessameparty
```

```
dim(model.matrix(model0, data=cm))
```

```
## [1] 425653       4
```

In contrast, the model matrix of specification (1) is of dimensions 425653×221, and the model matrix of specification (2) even of 425653×6929.

```
model1 <-
  yes ~ pctsumyessameschool +
        pctsumyessamestate +
        pctsumyessameparty +
        factor(congress) +
        factor(id) -1
mm1 <- model.matrix(model1, data=cm)
dim(mm1)
```

```
## [1] 425653     168
```

Using OLS to estimate such a model thus involves the computation of a very large matrix inversion (because $\hat{\beta}_{OLS} = (\mathbf{X}^\top\mathbf{X})^{-1}\mathbf{X}^\top\mathbf{y}$). In addition, the model matrix for specification 2 is about 22GB, which might further slow down the computer due to a lack of physical memory or even crash the R session altogether.

In order to set a point of reference, we first estimate specification (1) with standard OLS.

```
# fit specification (1)
runtime <- system.time(fit1 <- lm(data = cm, formula = model1))
coeftest(fit1)[2:4,]
```

```
##                       Estimate Std. Error t value
## pctsumyessamestate     0.11861   0.001085 109.275
## pctsumyessameparty     0.92640   0.001397 662.910
## factor(congress)101   -0.01458   0.006429  -2.269
##                       Pr(>|t|)
## pctsumyessamestate      0.0000
## pctsumyessameparty      0.0000
## factor(congress)101     0.0233
```

```
# median amount of time needed for estimation
runtime[3]
```

```
## elapsed
##   6.678
```

As expected, this takes quite some time to compute. However, there is an alternative approach to estimating such models that substantially reduces the computational burden by "sweeping out the fixed effects dummies". In the simple case of only one fixed effect variable (e.g., only individual fixed effects), the trick is called "within transformation" or "demeaning" and is quite simple to implement. For each of the categories in the fixed effect variable, compute the mean of the covariate and subtract the mean from the covariate's value.

```
# illustration of within transformation for the senator fixed effects
cm_within <-
  with(cm, data.table(yes = yes - ave(yes, id),
                      pctsumyessameschool = pctsumyessameschool -
                        ave(pctsumyessameschool, id),
                      pctsumyessamestate = pctsumyessamestate -
                        ave(pctsumyessamestate, id),
                      pctsumyessameparty = pctsumyessameparty -
                        ave(pctsumyessameparty, id)
                      ))
```

```
# comparison of dummy fixed effects estimator and within estimator
dummy_time <- system.time(fit_dummy <-
            lm(yes ~ pctsumyessameschool +
                pctsumyessamestate +
                pctsumyessameparty +
                factor(id) -1, data = cm
                ))
within_time <- system.time(fit_within <-
                        lm(yes ~ pctsumyessameschool +
                            pctsumyessamestate +
                            pctsumyessameparty -1,
                            data = cm_within))
# computation time comparison
as.numeric(within_time[3])/as.numeric(dummy_time[3])
```

```
## [1] 0.008582
```

```
# comparison of estimates
coeftest(fit_dummy)[1:3,]
```

```
##                        Estimate Std. Error t value
## pctsumyessameschool    0.04424    0.001352   32.73
## pctsumyessamestate     0.11864    0.001085  109.30
## pctsumyessameparty     0.92615    0.001397  662.93
##                          Pr(>|t|)
## pctsumyessameschool    1.205e-234
## pctsumyessamestate     0.000e+00
## pctsumyessameparty     0.000e+00
```

```
coeftest(fit_within)
```

```
##
## t test of coefficients:
##
##                        Estimate Std. Error t value
## pctsumyessameschool    0.04424     0.00135    32.7
## pctsumyessamestate     0.11864     0.00109   109.3
## pctsumyessameparty     0.92615     0.00140   663.0
##                         Pr(>|t|)
## pctsumyessameschool      <2e-16 ***
## pctsumyessamestate       <2e-16 ***
## pctsumyessameparty       <2e-16 ***
## ---
## Signif. codes:
## 0 '***' 0.001 '**' 0.01 '*' 0.05 '.' 0.1 ' ' 1
```

Unfortunately, we cannot simply apply the same procedure in a specification with several fixed effects variables. However, Gaure (2013b) provides a generalization of the linear within-estimator to several fixed effects variables. This method is implemented in the lfe package (Gaure, 2013a). With this package, we can easily estimate both fixed-effect specifications (as well as the corresponding cluster-robust standard errors) in order to replicate the original results by Cohen and Malloy (2014).

```
library(lfe)
```

```
# model and clustered SE specifications
model1 <- yes ~ pctsumyessameschool +
```

```
                    pctsumyessamestate +
                    pctsumyessameparty |congress+id|0|id
model2 <- yes ~ pctsumyessameschool +
                    pctsumyessamestate +
                    pctsumyessameparty |congress_session_votenumber+id|0|id


# estimation
fit1 <- felm(model1, data=cm)
fit2 <- felm(model2, data=cm)
```

Finally we can display the regression table.

```
stargazer::stargazer(fit1,fit2,
                     type="text",
                     dep.var.labels = "Vote (yes/no)",
                     covariate.labels = c("School Connected Votes",
                                          "State Votes",
                                          "Party Votes"),
                     keep.stat = c("adj.rsq", "n"))


##
## =====================================================
##                          Dependent variable:
##                     -----------------------------
##                              Vote (yes/no)
##                          (1)             (2)
## ---------------------------------------------------
## School Connected Votes   0.045***        0.052***
##                          (0.016)         (0.016)
##
## State Votes              0.119***        0.122***
##                          (0.013)         (0.012)
##
## Party Votes              0.926***        0.945***
##                          (0.022)         (0.024)
##
## -------------------------------------------------
## Observations             425,653         425,653
## Adjusted R2              0.641           0.641
## =====================================================
## Note:                    *p<0.1; **p<0.05; ***p<0.01
```

12.2   Case study: Loops, memory, and vectorization

We first read the `economics` dataset into R and extend it by duplicating its rows to get a slightly larger dataset (this step can easily be adapted to create a very large dataset).

```
# read dataset into R
economics <- read.csv("data/economics.csv")
# have a look at the data
head(economics, 2)
```

```
##          date   pce     pop psavert uempmed unemploy
## 1 1967-07-01 507.4 198712    12.5     4.5     2944
## 2 1967-08-01 510.5 198911    12.5     4.7     2945
```

```
# create a 'large' dataset out of this
for (i in 1:3) {
     economics <- rbind(economics, economics)
}
dim(economics)
```

```
## [1] 4592    6
```

The goal of this code example is to compute real personal consumption expenditures, assuming that `pce` in the `economics` dataset provides nominal personal consumption expenditures. Thus, we divide each value in the vector `pce` by a deflator `1.05`.

12.2.1   Naïve approach (ignorant of R)

The first approach we take is based on a simple `for` loop. In each iteration one element in `pce` is divided by the `deflator`, and the resulting value is stored as a new element in the vector `pce_real`.

```
# Naïve approach (ignorant of R)
deflator <- 1.05 # define deflator
# iterate through each observation
pce_real <- c()
n_obs <- length(economics$pce)
for (i in 1:n_obs) {
```

```
    pce_real <- c(pce_real, economics$pce[i]/deflator)
}
```

```
# look at the result
head(pce_real, 2)
```

```
## [1] 483.2 486.2
```

How long does it take?

```
# Naïve approach (ignorant of R)
deflator <- 1.05 # define deflator
# iterate through each observation
pce_real <- list()
n_obs <- length(economics$pce)
time_elapsed <-
    system.time(
        for (i in 1:n_obs) {
            pce_real <- c(pce_real, economics$pce[i]/deflator)
})
```

```
time_elapsed
```

```
##    user  system elapsed
##   0.105   0.004   0.110
```

Assuming a linear time algorithm ($O(n)$), we need that much time for one additional row of data:

```
time_per_row <- time_elapsed[3]/n_obs
time_per_row
```

```
##   elapsed
## 2.395e-05
```

If we are dealing with Big Data, say 100 million rows, that is

```
# in seconds
(time_per_row*100^4)
```

```
## elapsed
##    2395
```

```
# in minutes
(time_per_row*100^4)/60
```

```
## elapsed
##   39.92
```

```
# in hours
(time_per_row*100^4)/60^2
```

```
## elapsed
##   0.6654
```

Can we improve this?

12.2.2 Improvement 1: Pre-allocation of memory

In the naïve approach taken above, each iteration of the loop causes R to re-allocate memory because the number of elements in the vector `pce_element` is changing. In simple terms, this means that R needs to execute more steps in each iteration. We can improve this with a simple trick by initiating the vector to the right size to begin with (filled with NA values).

```
# Improve memory allocation (still somewhat ignorant of R)
deflator <- 1.05 # define deflator
n_obs <- length(economics$pce)
# allocate memory beforehand
# Initialize the vector to the right size
pce_real <- rep(NA, n_obs)
# iterate through each observation
time_elapsed <-
    system.time(
        for (i in 1:n_obs) {
            pce_real[i] <- economics$pce[i]/deflator
})
```

Let's see if this helped to make the code faster.

```
time_per_row <- time_elapsed[3]/n_obs
time_per_row
```

```
##    elapsed
## 1.742e-06
```

Again, we can extrapolate (approximately) the computation time, assuming the dataset had millions of rows.

```
# in seconds
(time_per_row*100^4)
```

```
## elapsed
##   174.2
```

```
# in minutes
(time_per_row*100^4)/60
```

```
## elapsed
##   2.904
```

```
# in hours
(time_per_row*100^4)/60^2
```

```
## elapsed
## 0.04839
```

This looks much better, but we can do even better.

12.2.3   Improvement 2: Exploit vectorization

In this approach, we exploit the fact that in R, 'everything is a vector' and that many of the basic R functions (such as math operators) are *vectorized*. In simple terms, this means that a vectorized operation is implemented in such a way that it can take advantage of the similarity of each of the vector's elements. That is, R only has to figure out once how to apply a given function to a vector element in order to apply it to all elements of the vector. In a simple loop, R has to go through the same 'preparatory' steps again and again in each iteration; this is time-intensive.

In this example, we specifically exploit that the division operator / is actually a vectorized function. Thus, the division by our `deflator` is applied to each element of `economics$pce`.

```
# Do it 'the R way'
deflator <- 1.05 # define deflator
# Exploit R's vectorization
time_elapsed <-
    system.time(
```

```
        pce_real <- economics$pce/deflator
              )
# same result
head(pce_real, 2)
```

```
## [1] 483.2 486.2
```

Now this is much faster. In fact, `system.time()` is not precise enough to capture the time elapsed. In order to measure the improvement, we use `microbenchmark::microbenchmark()` to measure the elapsed time in microseconds (millionth of a second).

```
library(microbenchmark)
# measure elapsed time in microseconds (avg.)
time_elapsed <-
  summary(microbenchmark(pce_real <- economics$pce/deflator))$mean
# per row (in sec)
time_per_row <- (time_elapsed/n_obs)/10^6
```

Now we get a more precise picture regarding the improvement due to vectorization:

```
# in seconds
(time_per_row*100^4)
```

```
## [1] 0.1687
```

```
# in minutes
(time_per_row*100^4)/60
```

```
## [1] 0.002811
```

```
# in hours
(time_per_row*100^4)/60^2
```

```
## [1] 4.685e-05
```

12.3   Case study: Bootstrapping and parallel processing

In this example, we estimate a simple regression model that aims to assess racial discrimination in the context of police stops.[2] The example is based on the 'Minneapolis Police Department 2017 Stop Dataset', containing data on nearly all stops made by the Minneapolis Police Department for the year 2017.

We start by importing the data into R.

```
url <-
"https://vincentarelbundock.github.io/Rdatasets/csv/carData/MplsStops.csv"
stopdata <- data.table::fread(url)
```

We specify a simple linear probability model that aims to test whether a person identified as 'white' is less likely to have their vehicle searched when stopped by the police. In order to take into account level differences between different police precincts, we add precinct indicators to the regression specification.

First, let's remove observations with missing entries (NA) and code our main explanatory variable and the dependent variable.

```
# remove incomplete obs
stopdata <- na.omit(stopdata)
# code dependent var
stopdata$vsearch <- 0
stopdata$vsearch[stopdata$vehicleSearch=="YES"] <- 1
# code explanatory var
stopdata$white <- 0
stopdata$white[stopdata$race=="White"] <- 1
```

We specify our baseline model as follows.

```
model <- vsearch ~ white + factor(policePrecinct)
```

and estimate the linear probability model via OLS (the lm function).

---

[2]Note that this example aims to illustrate a point about computation in an applied econometrics context. It does not make any argument whatsoever about identification or the broader research question.

```
fit <- lm(model, stopdata)
summary(fit)
```

```
##
## Call:
## lm(formula = model, data = stopdata)
##
## Residuals:
##     Min      1Q  Median      3Q     Max
## -0.1394 -0.0633 -0.0547 -0.0423  0.9773
##
## Coefficients:
##                          Estimate Std. Error t value
## (Intercept)               0.05473    0.00515   10.62
## white                    -0.01955    0.00446   -4.38
## factor(policePrecinct)2   0.00856    0.00676    1.27
## factor(policePrecinct)3   0.00341    0.00648    0.53
## factor(policePrecinct)4   0.08464    0.00623   13.58
## factor(policePrecinct)5  -0.01246    0.00637   -1.96
##                          Pr(>|t|)
## (Intercept)               < 2e-16 ***
## white                     1.2e-05 ***
## factor(policePrecinct)2      0.21
## factor(policePrecinct)3      0.60
## factor(policePrecinct)4   < 2e-16 ***
## factor(policePrecinct)5      0.05 .
## ---
## Signif. codes:
## 0 '***' 0.001 '**' 0.01 '*' 0.05 '.' 0.1 ' ' 1
##
## Residual standard error: 0.254 on 19078 degrees of freedom
## Multiple R-squared:  0.025,  Adjusted R-squared:  0.0248
## F-statistic: 97.9 on 5 and 19078 DF,  p-value: <2e-16
```

A potential problem with this approach (and there might be many more in this simple example) is that observations stemming from different police precincts might be correlated over time. If that is the case, we likely underestimate the coefficient's standard errors. There is a standard approach to computing estimates for so-called *cluster-robust* standard errors, which would take the problem of correlation over time within clusters into consideration (and deliver a more conservative estimate of the SEs). However, this approach only works well if the number of clusters in the data is roughly 50 or more. Here we only have five.

The alternative approach is to compute bootstrapped clustered standard errors. That is, we apply the bootstrap resampling procedure[3] at the cluster level. Specifically, we draw B samples (with replacement), estimate and record the coefficient vector for each bootstrap-sample, and then estimate SE_{boot} based on the standard deviation of all respective estimated coefficient values.

```r
# load packages
library(data.table)
# set the 'seed' for random numbers (makes the example reproducible)
set.seed(2)

# set number of bootstrap iterations
B <- 10
# get selection of precincts
precincts <- unique(stopdata$policePrecinct)
# container for coefficients
boot_coefs <- matrix(NA, nrow = B, ncol = 2)
# draw bootstrap samples, estimate model for each sample
for (i in 1:B) {

    # draw sample of precincts (cluster level)
    precincts_i <- base::sample(precincts, size = 5, replace = TRUE)
    # get observations
    bs_i <-
        lapply(precincts_i, function(x){
            stopdata[stopdata$policePrecinct==x,]
    } )
    bs_i <- rbindlist(bs_i)

    # estimate model and record coefficients
    boot_coefs[i,] <- coef(lm(model, bs_i))[1:2] # ignore FE-coefficients
}
```

Finally, let's compute SE_{boot}.

```r
se_boot <- apply(boot_coefs,
                 MARGIN = 2,
```

---

[3]https://en.wikipedia.org/wiki/Bootstrapping_(statistics)

```
                        FUN = sd)
se_boot
```

```
## [1] 0.004043 0.004690
```

Note that even with a very small B, computing SE_{boot} takes some time to compute. When setting B to over 500, computation time will be substantial. Also note that running this code hardly uses up more memory than the very simple approach without bootstrapping (after all, in each bootstrap iteration the dataset used to estimate the model is approximately the same size as the original dataset). There is little we can do to improve the script's performance regarding memory. However, we can tell R how to allocate CPU resources more efficiently to handle that many regression estimates.

In particular, we can make use of the fact that most modern computing environments (such as a laptop) have CPUs with several *cores*. We can exploit this fact by instructing the computer to run the computations *in parallel* (simultaneously computing on several cores). The following code is a parallel implementation of our bootstrap procedure that does exactly that.

```
# load packages for parallel processing
library(doSNOW)
# get the number of cores available
ncores <- parallel::detectCores()
# set cores for parallel processing
ctemp <- makeCluster(ncores) #
registerDoSNOW(ctemp)



# set number of bootstrap iterations
B <- 10
# get selection of precincts
precincts <- unique(stopdata$policePrecinct)
# container for coefficients
boot_coefs <- matrix(NA, nrow = B, ncol = 2)

# bootstrapping in parallel
boot_coefs <-
    foreach(i = 1:B, .combine = rbind, .packages="data.table") %dopar% {
        # draw sample of precincts (cluster level)
        precincts_i <- base::sample(precincts, size = 5, replace = TRUE)
```

```
        # get observations
        bs_i <- lapply(precincts_i, function(x) {
          stopdata[stopdata$policePrecinct==x,]
        })
        bs_i <- rbindlist(bs_i)
        # estimate model and record coefficients
        coef(lm(model, bs_i))[1:2] # ignore FE-coefficients
    }
# be a good citizen and stop the snow clusters
stopCluster(cl = ctemp)
```

As a last step, we again compute SE_{boot}.

```
se_boot <- apply(boot_coefs,
                 MARGIN = 2,
                 FUN = sd)
se_boot
```

```
## (Intercept)       white
##     0.001332     0.005642
```

12.3.1 Parallelization with an EC2 instance

This short tutorial illustrates how to scale up the computation of clustered standard errors shown above by running it on an AWS EC2 instance. Note that there are a few things that we need to keep in mind to make the script run on an AWS EC2 instance in RStudio Server. First, our EC2 instance is a Linux machine. When running R on a Linux machine, there is an additional step to install R packages (at least for most of the packages): R packages need to be compiled before they can be installed. The command to install packages is exactly the same (install.packages()), and normally you only notice a slight difference in the output shown in the R console during installation (and the installation process takes a little longer than you are used to). Apart from that, using R via RStudio Server in the cloud looks/feels very similar if not identical to when using R/RStudio locally. For this step of the case study, first follow the instructions of how to set up an AWS EC2 instance with R/RStudio Server in Chapter 7. Then, open a browser window, log in to RStudio Server on the EC2 instance, and copy and paste the code below to a new R-file on the EC2 instance (note that you might have to install the data.table and doSNOW packages before running the code).

When executing the code below line-by-line, you will notice that essentially all parts of the script work exactly as on your local machine. This is one of the great advantages of running R/RStudio Server in the cloud. You can implement your entire data analysis locally (based on a small sample), test it locally, and then move it to the cloud and run it at a larger scale in exactly the same way (even with the same Graphical User Interface (GUI)).

```r
# install packages
install.packages("data.table")
install.packages("doSNOW")
# load packages
library(data.table)


# fetch the data
url <- "https://vincentarelbundock.github.io/Rdatasets/csv/carData/MplsStops.csv"
stopdata <- read.csv(url)
# remove incomplete obs
stopdata <- na.omit(stopdata)
# code dependent var
stopdata$vsearch <- 0
stopdata$vsearch[stopdata$vehicleSearch == "YES"] <- 1
# code explanatory var
stopdata$white <- 0
stopdata$white[stopdata$race == "White"] <- 1


# model fit
model <- vsearch ~ white + factor(policePrecinct)
fit <- lm(model, stopdata)
summary(fit)
# bootstrapping: normal approach set the 'seed' for random
# numbers (makes the example reproducible)
set.seed(2)
# set number of bootstrap iterations
B <- 50
# get selection of precincts
precincts <- unique(stopdata$policePrecinct)
# container for coefficients
boot_coefs <- matrix(NA, nrow = B, ncol = 2)
# draw bootstrap samples, estimate model for each sample
for (i in 1:B) {
  # draw sample of precincts (cluster level)
  precincts_i <- base::sample(precincts, size = 5, replace = TRUE)
  # get observations
```

```r
  bs_i <- lapply(precincts_i, function(x) {
    stopdata[stopdata$policePrecinct == x, ]
  })
  bs_i <- rbindlist(bs_i)
  # estimate model and record coefficients
  boot_coefs[i, ] <- coef(lm(model, bs_i))[1:2]   # ignore FE-coefficients
}

se_boot <- apply(boot_coefs, MARGIN = 2, FUN = sd)
se_boot
```

So far, we have only demonstrated that the simple implementation (non-parallel) works both locally and in the cloud. However, the real purpose of using an EC2 instance in this example is to make use of the fact that we can scale up our instance to have more CPU cores available for the parallel implementation of our bootstrap procedure. Recall that running the script below on our local machine will employ all cores available to us and compute the bootstrap resampling in parallel on all these cores. Exactly the same thing happens when running the code below on our simple t2.micro instance. However, this type of EC2 instance only has one core. You can check this when running the following line of code in RStudio Server (assuming the doSNOW package is installed and loaded): parallel::detectCores() .

When running the entire parallel implementation below, you will thus notice that it won't compute the bootstrap SE any faster than with the non-parallel version above. However, by simply initiating another EC2 type with more cores, we can distribute the workload across many CPU cores, using exactly the same R script.

```r
# bootstrapping: parallel approaach
# install.packages("doSNOW", "parallel")
# load packages for parallel processing
library(doSNOW)
# set cores for parallel processing
ncores <- parallel::detectCores()
ctemp <- makeCluster(ncores)
registerDoSNOW(ctemp)
# set number of bootstrap iterations
B <- 50
# get selection of precincts
precincts <- unique(stopdata$policePrecinct)
# container for coefficients
boot_coefs <- matrix(NA, nrow = B, ncol = 2)
```

```r
# bootstrapping in parallel
boot_coefs <-
  foreach(i = 1:B, .combine = rbind, .packages="data.table") %dopar% {
    # draw sample of precincts (cluster level)
    precincts_i <- base::sample(precincts, size = 5, replace = TRUE)
    # get observations
    bs_i <- lapply(precincts_i, function(x){
        stopdata[stopdata$policePrecinct==x,])
    }
    bs_i <- rbindlist(bs_i)

    # estimate model and record coefficients
    coef(lm(model, bs_i))[1:2] # ignore FE-coefficients
  }


# be a good citizen and stop the snow clusters
stopCluster(cl = ctemp)
# compute the bootstrapped standard errors
se_boot <- apply(boot_coefs,
                 MARGIN = 2,
                 FUN = sd)
```

13

Econometrics with GPUs

GPUs have been used for a while in computational economics (see Aldrich (2014) for an overview of early applications in economics). However, until recently most of the work building on GPUs in economics has focused on solving economic models numerically (see, e.g., Aldrich et al. (2011)) and more broadly on Monte Carlo simulation. In this chapter, we first look at very basic GPU computation with R before having a look at the nowadays most common application of GPUs in applied econometrics, machine learning with neural networks.

13.1 OLS on GPUs

In a first simple tutorial, we have a look at how GPUs can be used to speed up basic econometric functions, such as the implementation of the OLS estimator. To this end, we will build on the gpuR package introduced in Chapter 5. To keep the example code simple, we follow the same basic set-up to implement and test our own OLS estimator function as in Chapter 3. That is, we first generate a sample based on (pseudo-)random numbers. To this end, we first define the sample size parameters n (the number of observations in our pseudo-sample) and p (the number of variables describing each of these observations) and then initialize the dataset X.

```
set.seed(1)
# set parameter values
n <- 100000
p <- 4
# generate a design matrix (~ our 'dataset')
# with p variables and n observations
X <- matrix(rnorm(n*p, mean = 10), ncol = p)
# add column for intercept
#X <- cbind(rep(1, n), X)
```

Following exactly the same code as in Chapter 3, we can now define what the real linear model that we have in mind looks like and compute the output y of this model, given the input X.

```
# MC model
y <-  1.5*X[,1] + 4*X[,2] - 3.5*X[,3] + 0.5*X[,4] + rnorm(n)
```

Now we re-implement our `beta_ols` function from Chapter 3 such that the OLS estimation is run on our local GPU. Recall that when computing on the GPU, we have the choice between keeping the data objects that go into the computation in RAM, or we can transfer the corresponding objects to GPU memory (which will further speed up the GPU computation). In the implementation of our `beta_ols_gpu`, I have added a parameter that allows switching between these two approaches. While setting `gpu_memory=TRUE` is likely faster, it might fail due to a lack of GPU memory (in all common desktop and laptop computers, RAM will be substantially larger than the GPU's own memory). Hence, `gpu_memory` is set to `FALSE` by default.

```
beta_ols_gpu <-
    function(X, y, gpu_memory=FALSE) {
        require(gpuR)

        if (!gpu_memory){
            # point GPU to matrix (matrix stored in non-GPU memory)
            vclX <- vclMatrix(X, type = "float")
            vcly <- vclVector(y, type = "float")
            # compute cross products and inverse
            XXi <- solve(crossprod(vclX,vclX))
            Xy <- crossprod(vclX, vcly)
        } else {
            # point GPU to matrix (matrix stored in non-GPU memory)
            gpuX <- gpuMatrix(X, type = "float")
            gpuy <- gpuVector(y, type = "float")
            # compute cross products and inverse
            XXi <- solve(crossprod(gpuX,gpuX))
            Xy <- t(gpuX)  %*% gpuy
        }
        beta_hat <- as.vector(XXi  %*% Xy)
        return(beta_hat)
    }
```

Now we can verify whether the implemented GPU-run OLS estimator works as expected.

```
beta_ols_gpu(X,y)
```

```
## [1]  1.5037  3.9997  -3.5036  0.5003
```

```
beta_ols_gpu(X,y, gpu_memory = TRUE)
```

```
## [1]  1.5033  3.9991  -3.5029  0.5005
```

Note how the coefficient estimates are very close to the true values. We can rest assured that our implementation of a GPU-based OLS estimator works fairly well. Also note how simple the basic implementation of functions to compute matrix-based operations on the GPU is through the gpuR package.

13.2   A word of caution

From just comparing the number of threads of a modern CPU with the number of threads of a modern GPU, one might get the impression that parallelizable tasks should always be implemented for GPU computing. However, whether one approach or the other is faster can depend a lot on the overall task and the data at hand. Moreover, the parallel implementation of tasks can be done more or less well on either system. Really efficient parallel implementation of tasks can take a lot of coding time (particularly when done for GPUs).[1]

As it turns out, the GPU OLS implementation above is actually a good example of a potential pitfall. While, as demonstrated in Chapter 4, matrix operations per se are likely much faster on GPUs than CPUs, the simple beta_ols_gpu() function implemented above involves more than the simple matrix operations. The model matrix as well as the vector of the dependent variable first had to be prepared for these operations (either a pointer for the GPU to the object in RAM had to be created or the objects had to be transferred to GPU memory). Finally, the computed values need to be transferred back to a normal R-object (at least if we want to make the output consistent with our simple beta_ols() implementation from Chapter 3). All of these steps create an additional overhead in terms of computing time.[2] Depending on the problem at hand, this overhead resulting from preparatory steps before running the actual computations on the GPU might be dwarfed by the efficiency gain if the computing task is much more demanding then what is involved in OLS. The

---

[1]For a more detailed discussion of the relevant factors for well-designed parallelization (either on CPUs or GPUs), see Matloff (2015).

[2]You can easily verify this by comparing the performance of beta_ols() (the simple CPU-based implementation) with the here implemented beta_ols_gpu() through the bench::mark() function.

section on TensorFlow/Keras below points to exactly such a setting, where GPUs
are typically much faster than CPUs.

13.3   Higher-level interfaces for basic econometrics with GPUs

The CRAN Task View on High-Performance and Parallel Computing with R[3] lists
several projects that provide easy-to-use interfaces to canned implementations of
regression and machine learning algorithms running on GPUs. For example, the
`tfestimators` package provides an R interface to use the TensorFlow Estimators
framework by Cheng et al. (2017). The package provides various canned estimators
to be run on GPUs (through TensorFlow).[4] Note, however, that this framework is
only compatible with TensorFlow version 1. As we will build on the latest version
of TensorFlow (version 2) in the following example (and as most applications now
build on version 2), we will not go into details of how to work with `tfestimators`.
However, there are excellent vignettes provided with the package that help you get
started.[5]

13.4   TensorFlow/Keras example: Predict housing prices

The most common application of GPUs in modern econometrics is machine learning, in particular deep learning (a type of machine learning based on artificial neural networks). Training deep learning models can be very computationally intensive and to a great extent depends on tensor (matrix) multiplications. This is also an
area where you might come across highly parallelized computing based on GPUs
without even noticing it, as the now commonly used software to build and train
deep neural nets (TensorFlow[6]; Abadi et al. (2015), and the high-level Keras[7] API;
Chollet et al. (2015)) can easily be run on a CPU or GPU without any further config-
uration/preparation (apart from the initial installation of these programs). In this
chapter, we look at a simple example of using GPUs with Keras in the context of
predictive econometrics.

In this example we train a simple sequential model with two hidden layers to predict the median value of owner-occupied homes (in USD 1,000) in the Boston area

---

[3]https://cran.r-project.org/web/views/HighPerformanceComputing.html
[4]See https://cran.r-project.org/web/packages/tfestimators/vignettes/estimator_basics.html
for an introduction to the basic usage of the package.
[5]See https://cran.r-project.org/web/packages/tfestimators/.
[6]https://www.tensorflow.org/
[7]https://keras.io/

(data is from the 1970s). The original data and a detailed description can be found here: https://www.cs.toronto.edu/~delve/data/boston/bostonDetail.html. The example closely follows this Keras tutorial[8] published by RStudio. See RStudio's Keras installation guide[9] for how to install Keras (and TensorFlow) and the corresponding R package `keras` (Allaire and Chollet, 2022).[10] While the purpose of the example here is to demonstrate a typical (but very simple!) use case of GPUs in machine learning, the same code should also run on a normal machine (without using GPUs) with a default installation of Keras.

Apart from `keras`, we load packages to prepare the data and visualize the output. Via `dataset_boston_housing()`, we load the dataset (shipped with the Keras installation) in the format preferred by the `keras` library.

```
# load packages
library(keras)
library(tibble)
library(ggplot2)
library(tfdatasets)
# load data
boston_housing <- dataset_boston_housing()
str(boston_housing)
```

```
## List of 2
##  $ train:List of 2
##   ..$ x: num [1:404, 1:13] 1.2325 0.0218 4.8982 0.0396 3.6931 ...
##   ..$ y: num [1:404(1d)] 15.2 42.3 50 21.1 17.7 18.5 11.3 15.6 15.6 14.4 ...
##  $ test :List of 2
##   ..$ x: num [1:102, 1:13] 18.0846 0.1233 0.055 1.2735 0.0715 ...
##   ..$ y: num [1:102(1d)] 7.2 18.8 19 27 22.2 24.5 31.2 22.9 20.5 23.2 ...
```

13.4.1 Data preparation

In a first step, we split the data into a training set and a test set. The latter is used to monitor the out-of-sample performance of the model fit. Testing the validity of an estimated model by looking at how it performs out-of-sample is of particular relevance when working with (deep) neural networks, as they can easily lead to overfitting. Validity checks based on the test sample are, therefore, often an integral part of modeling with TensorFlow/Keras.

---

[8]https://keras.rstudio.com/articles/tutorial_basic_regression.html#the-boston-housing-prices-dataset

[9]https://keras.rstudio.com/index.html

[10]This might involve the installation of additional packages and software outside the R environment. The following examples were run with TensorFlow version `tensorflow_gpu-2.9.3`.

```
# assign training and test data/labels
c(train_data, train_labels) %<-% boston_housing$train
c(test_data, test_labels) %<-% boston_housing$test
```

In order to better understand and interpret the dataset, we add the original variable names and convert it to a `tibble`.

```
library(dplyr)

column_names <- c('CRIM', 'ZN', 'INDUS', 'CHAS', 'NOX', 'RM', 'AGE',
                  'DIS', 'RAD', 'TAX', 'PTRATIO', 'B', 'LSTAT')

train_df <- train_data %>%
  as_tibble(.name_repair = "minimal") %>%
  setNames(column_names) %>%
  mutate(label = train_labels)

test_df <- test_data %>%
  as_tibble(.name_repair = "minimal") %>%
  setNames(column_names) %>%
  mutate(label = test_labels)
```

Next, we have a close look at the data. Note the usage of the term 'label' for what is usually called the 'dependent variable' in econometrics.[11] As the aim of the exercise is to predict median prices of homes, the output of the model will be a continuous value ('labels').

```
# check training data dimensions and content
dim(train_df)
```

```
## [1] 404  14
```

```
head(train_df)
```

```
## # A tibble: 6 x 14
##     CRIM    ZN INDUS  CHAS   NOX    RM   AGE   DIS
##    <dbl> <dbl> <dbl> <dbl> <dbl> <dbl> <dbl> <dbl>
```

---

[11]Typical textbook examples in machine learning deal with classification (e.g., a logit model), while in microeconometrics the typical example is usually a linear model (continuous dependent variable).

```
## 1 1.23      0    8.14    0 0.538  6.14  91.7  3.98
## 2 0.0218  82.5   2.03    0 0.415  7.61  15.7  6.27
## 3 4.90      0   18.1     0 0.631  4.97 100    1.33
## 4 0.0396    0    5.19    0 0.515  6.04  34.5  5.99
## 5 3.69      0   18.1     0 0.713  6.38  88.4  2.57
## 6 0.284     0    7.38    0 0.493  5.71  74.3  4.72
## # i 6 more variables: RAD <dbl>, TAX <dbl>,
## #   PTRATIO <dbl>, B <dbl>, LSTAT <dbl>,
## #   label <dbl[1d]>
```

As the dataset contains variables ranging from per capita crime rate to indicators for highway access, the variables are obviously measured in different units and hence displayed on different scales. This is not a problem per se for the fitting procedure. However, fitting is more efficient when all features (variables) are normalized.

```
spec <- feature_spec(train_df, label ~ . ) %>%
  step_numeric_column(all_numeric(), normalizer_fn = scaler_standard()) %>%
  fit()
```

13.4.2 Model specification

We specify the model as a linear stack of layers, the input (all 13 explanatory variables), two densely connected hidden layers (each with a 64-dimensional output space), and finally the one-dimensional output layer (the 'dependent variable').

```
# Create the model
# model specification
input <- layer_input_from_dataset(train_df %>% select(-label))

output <- input %>%
  layer_dense_features(dense_features(spec)) %>%
  layer_dense(units = 64, activation = "relu") %>%
  layer_dense(units = 64, activation = "relu") %>%
  layer_dense(units = 1)

model <- keras_model(input, output)
```

In order to fit the model, we first have to compile it (configure it for training). At this step we set the configuration parameters that will guide the training/optimization procedure. We use the mean squared errors loss function (mse) typically used for regressions, and we chose the RMSProp[12] optimizer to find the minimum loss.

```
# compile the model
model %>%
  compile(
    loss = "mse",
    optimizer = optimizer_rmsprop(),
    metrics = list("mean_absolute_error")
  )
```

Now we can get a summary of the model we are about to fit to the data.

```
# get a summary of the model
model
```

13.4.3   Training and prediction

Given the relatively simple model and small dataset, we set the maximum number of epochs to 500.

```
# Set max. number of epochs
epochs <- 500
```

Finally, we fit the model while preserving the training history, and visualize the training progress.

```
# Fit the model and store training stats
history <- model %>% fit(
  x = train_df %>% select(-label),
  y = train_df$label,
  epochs = epochs,
  validation_split = 0.2,
  verbose = 0
)
plot(history)
```

---

[12]http://www.cs.toronto.edu/~tijmen/csc321/slides/lecture_slides_lec6.pdf

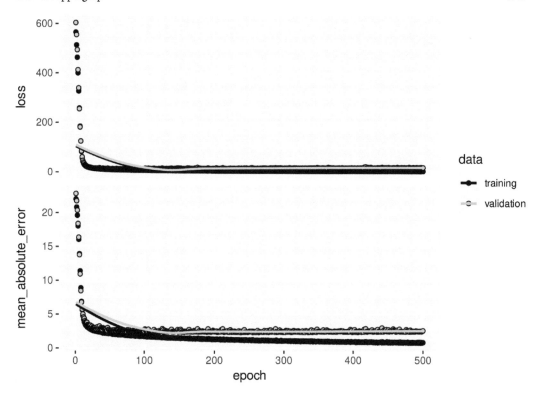

13.5 Wrapping up

- gpuR provides a straightforward interface for applied econometrics run on GPUs. While working with gpuR, be aware of the necessary computational overhead to run commands on the GPU via this interface. For example, implementing the OLS estimator with gpuR is a good exercise but does not really pay off in terms of performance.
- There are several ongoing projects in the R world to bring GPU computation closer to basic data analytics tasks, providing high-level interfaces to work with GPUs (see the CRAN Task View on High-Performance and Parallel Computing with R[13] for some of those).
- A typical application of GPU computation in applied econometrics is the training of neural nets, particularly deep neural nets (deep learning). The keras and tensorflow packages provide excellent R interfaces to work with the deep learning libraries TensorFlow and Keras. Both of those libraries are implemented to directly work with GPUs.

---

[13]https://cran.r-project.org/web/views/HighPerformanceComputing.html

14

Regression Analysis and Categorization with Spark and R

Regression analysis, particularly simple linear regression (OLS), is the backbone of applied econometrics. As discussed in previous chapters, regression analysis can be computationally very intensive with a dataset of many observations and variables, as it involves matrix operations on a very large model matrix. Chapter 12 discusses in one case study the special case of a large model matrix due to fixed-effects dummy variables. In this chapter, we first look at a generally applicable approach for estimating linear regression models with large datasets (when the model matrix cannot be held in RAM). Building on the same `sparklyr` framework (Luraschi et al., 2022) as for the simple linear regression case, we then look at classification models, such as logit and random forest. Finally, we look at how regression analysis and machine learning tasks can be organized in machine learning pipelines to be run, stored/reloaded, and updated flexibly.

14.1 Simple linear regression analysis

Suppose we want to conduct a correlation study of what factors are associated with longer or shorter arrival delays in air travel. Via its built-in 'MLib' library, Spark provides several high-level functions to conduct regression analyses. When calling these functions via `sparklyr` (or `SparkR`), their usage is actually very similar to the usual R packages/functions commonly used to run regressions in R.

As a simple point of reference, we first estimate a linear model with the usual R approach (all computed in the R environment). First, we load the data as a common `data.table`. We could also convert a copy of the entire `SparkDataFrame` object to a `data.frame` or `data.table` and get essentially the same outcome. However, collecting the data from the RDD structure would take much longer than parsing the CSV with `fread`. In addition, we only import the first 300 rows. Running regression analysis with relatively large datasets in Spark on a small local machine might fail or be rather slow.[1]

---

[1] Again, it is important to keep in mind that running Spark on a small local machine is only optimal for learning and testing code (based on relatively small samples). The whole framework is optimized to be run on cluster computers.

```
# flights_r <- collect(flights) # very slow!
flights_r <- data.table::fread("data/flights.csv", nrows = 300)
```

Now we run a simple linear regression (OLS) and show the summary output.

```
# specify the linear model
model1 <- arr_delay ~ dep_delay + distance
# fit the model with OLS
fit1 <- lm(model1, flights_r)
# compute t-tests etc.
summary(fit1)
```

```
##
## Call:
## lm(formula = model1, data = flights_r)
##
## Residuals:
##    Min     1Q Median     3Q    Max
## -42.39  -9.96  -1.91   9.87  48.02
##
## Coefficients:
##               Estimate Std. Error t value Pr(>|t|)
## (Intercept) -0.182662   1.676560   -0.11     0.91
## dep_delay    0.989553   0.017282   57.26   <2e-16 ***
## distance     0.000114   0.001239    0.09     0.93
## ---
## Signif. codes:
## 0 '***' 0.001 '**' 0.01 '*' 0.05 '.' 0.1 ' ' 1
##
## Residual standard error: 15.5 on 297 degrees of freedom
## Multiple R-squared:  0.917,  Adjusted R-squared:  0.917
## F-statistic: 1.65e+03 on 2 and 297 DF,  p-value: <2e-16
```

Now we aim to compute essentially the same model estimate in sparklyr.[2] In order to use Spark via the sparklyr package, we need to first load the package and establish a connection with Spark (similar to SparkR::sparkR.session()).

---

[2]Most regression models commonly used in traditional applied econometrics are provided in some form in sparklyr or SparkR. See the package documentation for more details.

```
library(sparklyr)

# connect with default configuration
sc <- spark_connect(master="local")
```

We then copy the data.table `flights_r` (previously loaded into our R session) to Spark. Again, working on a normal laptop this seems trivial, but the exact same command would allow us (when connected with Spark on a cluster computer in the cloud) to properly load and distribute the data.table on the cluster. Finally, we then fit the model with `ml_linear_regression()` and compute.

```
# load data to spark
flights_spark <- copy_to(sc, flights_r, "flights_spark")
# fit the model
fit1_spark <- ml_linear_regression(flights_spark, formula = model1)
# compute summary stats
summary(fit1_spark)


Deviance Residuals:
    Min      1Q  Median      3Q     Max
-42.386  -9.965  -1.911   9.866  48.024


Coefficients:
  (Intercept)      dep_delay       distance
-0.1826622687   0.9895529018   0.0001139616


R-Squared: 0.9172
Root Mean Squared Error: 15.42
```

Alternatively, we can use the `spark_apply()` function to run the regression analysis in R via the original R `lm()` function.[3]

```
# fit the model
spark_apply(flights_spark,
            function(df){
                broom::tidy(lm(arr_delay ~ dep_delay + distance, df))},
            names = c("term",
                      "estimate",
                      "std.error",
```

---

[3]Note, though, that this approach might take longer.

```
                    "statistic",
                    "p.value")
   )
```

```
# Source: spark<?> [?? x 5]
  term           estimate std.error statistic  p.value
  <chr>             <dbl>     <dbl>     <dbl>    <dbl>
1 (Intercept) -0.183         1.68      -0.109 9.13e-  1
2 dep_delay     0.990        0.0173    57.3   1.63e-162
3 distance      0.000114     0.00124    0.0920 9.27e-  1
```

Finally, the `parsnip` package (Kuhn and Vaughan, 2022) (together with the `tidymodels` package; Kuhn and Wickham (2020)) provides a simple interface to run the same model (or similar specifications) on different "engines" (estimators/fitting algorithms), and several of the `parsnip` models are also supported in `sparklyr`. This significantly facilitates the transition from local testing (with a small subset of the data) to running the estimation on the entire dataset on spark.

```
library(tidymodels)
library(parsnip)

# simple local linear regression example from above
# via tidymodels/parsnip
fit1 <- fit(linear_reg(engine="lm"), model1, data=flights_r)
tidy(fit1)
```

```
# A tibble: 3 × 5
  term           estimate std.error statistic  p.value
  <chr>             <dbl>     <dbl>     <dbl>    <dbl>
1 (Intercept) -0.183         1.68      -0.109 9.13e-  1
2 dep_delay     0.990        0.0173    57.3   1.63e-162
3 distance      0.000114     0.00124    0.0920 9.27e-  1
```

```
# run the same on Spark
fit1_spark <- fit(linear_reg(engine="spark"), model1, data=flights_spark)
tidy(fit1_spark)
```

```
# A tibble: 3 × 5
  term           estimate std.error statistic  p.value
  <chr>             <dbl>     <dbl>     <dbl>    <dbl>
1 (Intercept) -0.183         1.68      -0.109 9.13e-  1
```

```
2 dep_delay    0.990      0.0173    57.3    1.63e-162
3 distance     0.000114   0.00124    0.0920 9.27e-  1
```

We will further build on this interface in the next section where we look at different machine learning procedures for a classification problem.

14.2 Machine learning for classification

Building on `sparklyr`, `tidymodels`, and `parsnip`, we test a set of machine learning models on the classification problem discussed in Varian (2014), predicting Titanic survivors. The data for this exercise can be downloaded from here: http://doi.org/10.3886/E113925V1.

We import and prepare the data in R.

```
# load into R, select variables of interest, remove missing
titanic_r <- read.csv("data/titanic3.csv")
titanic_r <- na.omit(titanic_r[, c("survived",
                        "pclass",
                        "sex",
                        "age",
                        "sibsp",
                        "parch")])
titanic_r$survived <- ifelse(titanic_r$survived==1, "yes", "no")
```

In order to assess the performance of the classifiers later on, we split the sample into training and test datasets. We do so with the help of the `rsample` package (Frick et al., 2022), which provides a number of high-level functions to facilitate this kind of pre-processing.

```
library(rsample)
```

```
# split into training and test set
titanic_r <- initial_split(titanic_r)
ti_training <- training(titanic_r)
ti_testing <- testing(titanic_r)
```

For the training and assessment of the classifiers, we transfer the two datasets to the spark cluster.

```
# load data to spark
ti_training_spark <- copy_to(sc, ti_training, "ti_training_spark")
ti_testing_spark <- copy_to(sc, ti_testing, "ti_testing_spark")
```

Now we can set up a 'horse race' between different ML approaches to find the best performing model. Overall, we will consider the following models/algorithms:

- Logistic regression
- Boosted trees
- Random forest

```
# models to be used
models <- list(logit=logistic_reg(engine="spark", mode = "classification"),
               btree=boost_tree(engine = "spark", mode = "classification"),
               rforest=rand_forest(engine = "spark", mode = "classification"))
# train/fit the models
fits <- lapply(models, fit, formula=survived~., data=ti_training_spark)
```

The fitted models (trained algorithms) can now be assessed with the help of the test dataset. To this end, we use the high-level accuracy function provided in the yardstick package (Kuhn et al., 2022) to compute the accuracy of the fitted models. We proceed in three steps. First, we use the fitted models to predict the outcomes (we classify cases into survived/did not survive) of the *test set*. Then we fetch the predictions from the Spark cluster, format the variables, and add the actual outcomes as an additional column.

```
# run predictions
predictions <- lapply(fits, predict, new_data=ti_testing_spark)
# fetch predictions from Spark, format, add actual outcomes
pred_outcomes <-
    lapply(1:length(predictions), function(i){
        x_r <- collect(predictions[[i]]) # load into local R environment
        x_r$pred_class <- as.factor(x_r$pred_class) # format for predictions
        x_r$survived <- as.factor(ti_testing$survived) # add true outcomes
        return(x_r)

})
```

Finally, we compute the accuracy of the models, stack the results, and display them (ordered from best-performing to worst-performing.)

```
acc <- lapply(pred_outcomes, accuracy, truth="survived", estimate="pred_class")
acc <- bind_rows(acc)
acc$model <- names(fits)
acc[order(acc$.estimate, decreasing = TRUE),]
```

```
# A tibble: 3 × 4
  .metric  .estimator .estimate model
  <chr>    <chr>          <dbl> <chr>
1 accuracy binary         0.817 rforest
2 accuracy binary         0.790 btree
3 accuracy binary         0.779 logit
```

In this simple example, all models perform similarly well. However, none of them really performs outstandingly. In a next step, we might want to learn about which variables are considered more or less important for the predictions. Here, the tidy() function is very useful. As long as the model types are comparable (here btree and rforest), tidy() delivers essentially the same type of summary for different models.

```
tidy(fits[["btree"]])
```

```
# A tibble: 5 × 2
  feature   importance
  <chr>          <dbl>
1 age            0.415
2 sex_male       0.223
3 pclass         0.143
4 sibsp          0.120
5 parch          0.0987
```

```
tidy(fits[["rforest"]])
```

```
# A tibble: 5 × 2
  feature   importance
  <chr>          <dbl>
1 sex_male       0.604
2 pclass         0.188
3 age            0.120
4 sibsp          0.0595
5 parch          0.0290
```

Finally, we clean up and disconnect from the Spark cluster.

```
spark_disconnect(sc)
```

14.3 Building machine learning pipelines with R and Spark

Spark provides a framework to implement machine learning pipelines called ML Pipelines[4], with the aim of facilitating the combination of various preparatory steps and ML algorithms into a pipeline/workflow. `sparklyr` provides a straightforward interface to ML Pipelines that allows implementing and testing the entire ML workflow in R and then easily deploying the final pipeline to a Spark cluster or more generally to the production environment. In the following example, we will revisit the e-commerce purchase prediction model (Google Analytics data from the Google Merchandise Shop) introduced in Chapter 1. That is, we want to prepare the Google Analytics data and then use lasso to find a set of important predictors for purchase decisions, all built into a machine learning pipeline.

14.3.1 Set up and data import

All of the key ingredients are provided in `sparklyr`. However, I recommend using the 'piping' syntax provided in `dplyr` (Wickham et al., 2023) to implement the machine learning pipeline. In this context, using this syntax is particularly helpful to make the code easy to read and understand.

```
# load packages
library(sparklyr)
library(dplyr)

# fix vars
INPUT_DATA <- "data/ga.csv"
```

Recall that the Google Analytics dataset is a small subset of the overall data generated by Google Analytics on a moderately sized e-commerce site. Hence, it makes perfect sense to first implement and test the pipeline locally (on a local Spark installation) before deploying it on an actual Spark cluster in the cloud. In a first step, we thus copy the imported data to the local Spark instance.

---

[4]https://spark.apache.org/docs/latest/ml-pipeline.html

```
# import to local R session, prepare raw data
ga <- na.omit(read.csv(INPUT_DATA))
#ga$purchase <- as.factor(ifelse(ga$purchase==1, "yes", "no"))
# connect to, and copy the data to the local cluster
sc <- spark_connect(master = "local")
ga_spark <- copy_to(sc, ga, "ga_spark", overwrite = TRUE)
```

14.3.2   Building the pipeline

The pipeline object is initialized via ml_pipeline(), in which we refer to the connection to the local Spark cluster. We then add the model specification (the formula) to the pipeline with ft_r_formula(). ft_r_formula essentially transforms the data in accordance with the common specification syntax in R (here: purchase ~ .). Among other things, this takes care of properly setting up the model matrix. Finally, we add the model via ml_logistic_regression(). We can set the penalization parameters via elastic_net_param (with alpha=1, we get the lasso).

```
# ml pipeline
ga_pipeline <-
    ml_pipeline(sc) %>%
    ft_string_indexer(input_col="city",
                      output_col="city_output",
                      handle_invalid = "skip") %>%
    ft_string_indexer(input_col="country",
                      output_col="country_output",
                      handle_invalid = "skip") %>%
    ft_string_indexer(input_col="source",
                      output_col="source_output",
                      handle_invalid = "skip") %>%
    ft_string_indexer(input_col="browser",
                      output_col="browser_output",
                      handle_invalid = "skip") %>%
    ft_r_formula(purchase ~ .) %>%
    ml_logistic_regression(elastic_net_param = list(alpha=1))
```

Finally, we create a cross-validator object to train the model with a k-fold cross-validation and fit the model. For the sake of the example, we use only a 30-fold cross validation (to be run in parallel on 8 cores).

```
# specify the hyperparameter grid
# (parameter values to be considered in optimization)
ga_params <- list(logistic_regression=list(max_iter=80))

# create the cross-validator object
set.seed(1)
cv_lasso <- ml_cross_validator(sc,
                    estimator=ga_pipeline,
                    estimator_param_maps = ga_params,
                    ml_binary_classification_evaluator(sc),
                    num_folds = 30,
                    parallelism = 8)

# train/fit the model
cv_lasso_fit <- ml_fit(cv_lasso, ga_spark)
# note: this takes several minutes to run on a local machine (1 node, 8 cores)
```

Finally, we can inspect and further process the results – in particular the model's performance.

```
# pipeline summary
# cv_lasso_fit
# average performance
cv_lasso_fit$avg_metrics_df
```

```
  areaUnderROC max_iter_1
1   0.8666304         80
```

Before closing the connection to the Spark cluster, we can save the entire pipeline to work further with it later on.

```
# save the entire pipeline/fit
ml_save(
  cv_lasso_fit,
  "ga_cv_lasso_fit",
  overwrite = TRUE
)
```

To reload the pipeline later on, run `ml_load(sc, "ga_cv_lasso_fit")`.

14.4 Wrapping up

The key take-aways from this chapter are:

- When running econometric analysis such as linear or logistic regressions with massive amounts of data, `sparklyr` provides all the basic functions you need.
- You can test your code on your local spark installation by connecting to the local 'cluster': `spark_connect(master="local")`. This allows you to test your entire regression analysis script locally (on a sub-sample) before running the exact same script via a connection to a large spark cluster on AWS EMR. To do so, simply connect to the cluster via `spark_connect(master = "yarn")` from RStudio server, following the setup introduced in Section 8.4.
- The `rsample` package provides easy-to-use high-level functions to split your dataset into training and test datasets: See `?initial_split`, `?training`, and `?testing`.
- The `parsnip` and `broom` packages (Robinson et al., 2022) provide a way to easily standardize regression output. This is very helpful if you want to verify your regression analysis implementation for Spark with the more familiar R regression frameworks such as `lm()`. For example, compare the standard R OLS output with the linear regression output computed on a Spark cluster: `fit(linear_reg(engine="lm"), model1, data=flights_r)` for R's standard OLS; `fit(linear_reg(engine="spark"), model1, data=flights_r)` for Spark.
- For more advanced users, `sparklyr` provides a straightforward way to efficiently implement entire Spark machine learning pipelines in an R script via `ml_pipeline(sc)` and the `dplyr`-style pipe operators `%>%`, including model specification, data preparation, and selection and specification of the estimator.

15

Large-scale Text Analysis with sparklyr

Text analysis/natural language processing (NLP) often involves rather large amounts of data and is particularly challenging for in-memory processing. sparklyr provides several easy-to-use functions to run some of the computationally most demanding text data handling on a Spark cluster. In this chapter we explore these functions and the corresponding workflows to do text analysis on an AWS EMR cluster running Spark. Thereby we focus on the first few key components of a modern NLP pipeline. Figure 15.1 presents an overview of the main components of such a pipeline.

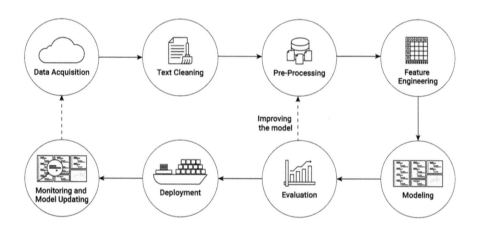

FIGURE 15.1: Illustration of an NLP (Natural Language Processing) pipeline.

Up until the deployment of an NLP model, all the steps involved constitute the typical workflow of economic research projects based on text data. Conveniently, all these first crucial steps of analyzing text data are covered in a few high-level functions provided in the sparklyr package. Implementing these steps and running them based on massive amounts of text data on an AWS EMR cluster are thus straightforward.

To get familiar with the basic syntax, the following subsection covers the first steps in such a pipeline based on a very simple text example.

15.1   Getting started: Import, pre-processing, and word count

The following example briefly guides the reader through some of the most common first steps when processing text data for NLP. In the code example, we process Friedrich Schiller's "Wilhelm Tell" (English edition; Project Gutenberg Book ID 2782), which we download from Project Gutenberg[1] by means of the gutenbergr package (Johnston and Robinson, 2022). The example can easily be extended to process many more books.

The example is set up to work straightforwardly on an AWS EMR cluster. However, given the relatively small amount of data processed here, you can also run it locally. If you want to run it on EMR, simply follow the steps in Section 6.4 to set up the cluster and log in to RStudio on the master node. The sparklyr package is already installed on EMR (if you use the bootstrap-script introduced in Section 6.4 for the set-up of the cluster), but other packages might still have to be installed.

We first load the packages and connect the RStudio session to the cluster (if you run this locally, use spark_connect(master="local")).

```r
# install additional packages
# install.packages("gutenbergr") # download book from Project Gutenberg
# install.packages("dplyr") # for the data preparatory steps


# load packages
library(sparklyr)
library(gutenbergr)
library(dplyr)


# fix vars
TELL <- "https://www.gutenberg.org/cache/epub/6788/pg6788.txt"



# connect rstudio session to cluster
sc <- spark_connect(master = "yarn")
```

We fetch the raw text of the book and copy it to the Spark cluster. Note that you can do this sequentially for many books without exhausting the master node's RAM and then further process the data on the cluster.

---

[1]https://www.gutenberg.org/

```
# Data gathering and preparation
# fetch Schiller's Tell, load to cluster
tmp_file <- tempfile()
download.file(TELL, tmp_file)
raw_text <- readLines(tmp_file)
tell <- data.frame(raw_text=raw_text)
tell_spark <- copy_to(sc, tell,
                      "tell_spark",
                      overwrite = TRUE)
```

The text data will be processed in a `SparkDataFrame` column behind the `tell_spark` object. First, we remove empty lines of text, select the column containing all the text, and then remove all non-numeric and non-alphabetical characters. The last step is an important text cleaning step as we want to avoid special characters being considered words or parts of words later on.

```
# data cleaning
tell_spark <- filter(tell_spark, raw_text!="")
tell_spark <- select(tell_spark, raw_text)
tell_spark <- mutate(tell_spark,
                     raw_text = regexp_replace(raw_text, "[^0-9a-zA-Z]+", " "))
```

Now we can split the lines of text in the column `raw_text` into individual words (sequences of characters separated by white space). To this end we can call a Spark feature transformation routine called tokenization, which essentially breaks text into individual terms. Specifically, each line of raw text in the column `raw_text` will be split into words. The overall result (stored in a new column specified with `output_col`), is then a nested list in which each word is an element of the corresponding line element.

```
# split into words
tell_spark <- ft_tokenizer(tell_spark,
                           input_col = "raw_text",
                           output_col = "words")
```

Now we can call another feature transformer called "stop words remover", which excludes all the stop words (words often occurring in a text but not carrying much information) from the nested word list.

```
# remove stop-words
tell_spark <- ft_stop_words_remover(tell_spark,
                                    input_col = "words",
                                    output_col = "words_wo_stop")
```

Finally, we combine all of the words in one vector and store the result in a new SparkDataFrame called "all_tell_words" (by calling compute()) and add some final cleaning steps.

```
# unnest words, combine in one row
all_tell_words <- mutate(tell_spark,
                word = explode(words_wo_stop))

# final cleaning
all_tell_words <- select(all_tell_words, word)
all_tell_words <- filter(all_tell_words, 2<nchar(word))
```

Based on this cleaned set of words, we can compute the word count for the entire book.

```
# get word count and store result in Spark memory
compute(count(all_tell_words, word), "wordcount_tell")

## # Source: spark<wordcount_tell> [?? x 2]
##     word            n
##     <chr>        <int>
##  1 language         2
##  2 martin           1
##  3 baron            8
##  4 nephew           3
##  5 hofe             3
##  6 reding          16
##  7 fisherman       39
##  8 baumgarten      32
##  9 hildegard        3
## 10 soldiers         4
## # i more rows
```

Finally, we can disconnect the R session from the Spark cluster

```
spark_disconnect(sc)
```

15.2   Tutorial: political slant

The tutorial below shows how to use `sparklyr` (in conjunction with AWS EMR) to run the entire raw text processing of academic research projects in economics. We will replicate the data preparation and computation of the *slant measure* for congressional speeches suggested by Gentzkow and Shapiro (2010) in the tutorial. To keep things simple, we'll use the data compiled by Gentzkow et al. (2019) and made available at https://data.stanford.edu/congress_text.

15.2.1   Data download and import

To begin, we download the corresponding zip-file to the EMR master node (if using EMR) or to your local RStudio working directory if using a local Spark installation. The unzipped folder contains text data from all speeches delivered from the 97th to the 114th US Congress, among other things. We will primarily work with the raw speeches text data, which is stored in files with the naming pattern `"speeches CONGRESS.txt,"` where `CONGRESS` is the number of the corresponding US Congress. To make things easier, we put all of the `speeches`-files in a subdirectory called `'speeches'`. The following section simply illustrates one method for downloading and rearranging the data files. The following code chunks require that all files containing the text of speeches be stored in `data/text/speeches` and all speaker information be stored in `data/text/speakers`.

```
# download and unzip the raw text data
URL <- "https://stacks.stanford.edu/file/druid:md374tz9962/hein-daily.zip"
PATH <- "data/hein-daily.zip"
system(paste0("curl ",
              URL,
              " > ",
              PATH,
              " && unzip ",
              PATH))
# move the speeches files
system("mkdir data/text/ && mkdir data/text/speeches")
system("mv hein-daily/speeches* data/text/speeches/")
# move the speaker files
```

```
system("mkdir data/text/speakers")
system("mv hein-daily/*SpeakerMap.txt data/text/speakers/")
```

In addition, we download an extra file in which the authors kept only valid phrases (after removing procedural phrases that often occur in congressional speeches but that do not contribute to finding partisan phrases). Thus we can later use this additional file to filter out invalid bigrams.[2]

```
# download and unzip procedural phrases data
URL_P <- "https://stacks.stanford.edu/file/druid:md374tz9962/vocabulary.zip"
PATH_P <- "data/vocabulary.zip"
system(paste0("curl ",
              URL_P,
              " > ",
              PATH_P,
              " && unzip ",
              PATH_P))
# move the procedural vocab file
system("mv vocabulary/vocab.txt data/text/")
```

We begin by loading the corresponding packages, defining some fixed variables, and connecting the R session to the Spark cluster, using the same basic pipeline structure as in the previous section's introductory example. Typically, you would first test these steps on a local Spark installation before feeding in more data to process on a Spark cluster in the cloud. In the following example, we only process the congressional speeches from the 97th to the 114th US Congress. The original source provides data for almost the entire history of Congress (see https://data.stanford.edu/congress_text for details). Recall that for local tests/working with the local Spark installation, you can connect your R session with sc <- spark_connect(master = "local"). Since even the limited speeches set we work with locally is several GBs in size, we set the memory available to our local Spark node to 16GB. This can be done by fetching the config file via spark_config() and then setting the driver-memory accordingly before initializing the Spark connection with the adapted configuration object (see config = conf in the spark_connect() command).

Unlike in the simple introductory example above, the raw data is distributed in multiple files. By default, Spark expects to load data from multiple files in the same directory. Thus, we can use the spark_read_csv() function to specify where all of the raw data is located in order to read in all of the raw text data at once. The data in this example is essentially stored in CSV format, but the pipe symbol | is used to

---

[2]Invalid for the context of this study.

separate columns instead of the more common commas. By specifying `delimiter="|"`, we ensure that the data structure is correctly captured.

```
# LOAD TEXT DATA  -------------------

# load data
speeches <- spark_read_csv(sc,
                           name = "speeches",
                           path =  INPUT_PATH_SPEECHES,
                           delimiter = "|")
speakers <- spark_read_csv(sc,
                           name = "speakers",
                           path =  INPUT_PATH_SPEAKERS,
                           delimiter = "|")
```

15.2.2 Cleaning speeches data

The intermediate goal of the data preparation steps is to determine the number of bigrams per party. That is, we want to know how frequently members of a particular political party have used a two-word phrase. As a first step, we must combine the speeches and speaker data to obtain the party label per speech and then clean the raw text to extract words and create and count bigrams. Congressional speeches frequently include references to dates, bill numbers, years, and so on. This introduces a slew of tokens made up entirely of digits, single characters, and special characters. The cleaning steps that follow are intended to remove the majority of those.

```
# JOIN -------------------
speeches <-
    inner_join(speeches,
               speakers,
               by="speech_id") %>%
    filter(party %in% c("R", "D"), chamber=="H")  %>%
    mutate(congress=substr(speech_id, 1,3)) %>%
    select(speech_id, speech, party, congress)


# CLEANING ----------------
# clean text: numbers, letters (bill IDs, etc.
speeches <-
    mutate(speeches, speech = tolower(speech)) %>%
```

```
    mutate(speech = regexp_replace(speech,
                                    "[_\"\'():;,.!?\\-]",
                                    "")) %>%
    mutate(speech = regexp_replace(speech, "\\\\(.+\\\\)", " ")) %>%
    mutate(speech = regexp_replace(speech, "[0-9]+", " ")) %>%
    mutate(speech = regexp_replace(speech, "<[a-z]+>", " ")) %>%
    mutate(speech = regexp_replace(speech, "<\\w+>", " ")) %>%
    mutate(speech = regexp_replace(speech, "_", " ")) %>%
    mutate(speech = trimws(speech))
```

15.2.3   Create a bigrams count per party

Based on the cleaned text, we now split the text into words (tokenization), remove stopwords, and create a list of bigrams (2-word phrases). Finally, we unnest the bigram list and keep the party and bigram column. The resulting Spark table contains a row for each bigram mentioned in any of the speeches along with the information of whether the speech in which the bigram was mentioned was given by a Democrat or a Republican.

```
# TOKENIZATION, STOPWORDS REMOVAL, NGRAMS ----------------


# stopwords list
stop <- readLines("http://snowball.tartarus.org/algorithms/english/stop.txt")
stop <- trimws(gsub("\\|.*", "", stop))
stop <- stop[stop!=""]


# clean text: numbers, letters (bill IDs, etc.
bigrams <-
    ft_tokenizer(speeches, "speech", "words")  %>%
    ft_stop_words_remover("words", "words_wo_stop",
                          stop_words = stop )  %>%
    ft_ngram("words_wo_stop", "bigram_list", n=2)   %>%
    mutate(bigram=explode(bigram_list)) %>%
    mutate(bigram=trim(bigram)) %>%
    mutate(n_words=as.numeric(length(bigram) -
                              length(replace(bigram, ' ', '')) + 1)) %>%
    filter(3<nchar(bigram), 1<n_words) %>%
    select(party, congress, bigram)
```

Before counting the bigrams by party, we need an additional context-specific cleaning step in which we remove procedural phrases from the speech bigrams.

```
# load the procedural phrases list
valid_vocab <- spark_read_csv(sc,
                                path="data/text/vocab.txt",
                                name = "valid_vocab",
                                delimiter = "|",
                                header = FALSE)
# remove corresponding bigrams via anti-join
bigrams <- inner_join(bigrams, valid_vocab, by= c("bigram"="V1"))
```

15.2.4 Find "partisan" phrases

At this point, we have all pieces in place in order to compute the bigram count (how often a certain bigram was mentioned by a member of either party). As this is an important intermediate result, we evaluate the entire operation for all the data and cache it in Spark memory through `compute()`. Note that if you run this code on your local machine, it can take a while to process.

```
# BIGRAM COUNT PER PARTY ---------------
bigram_count <-
    count(bigrams, party, bigram, congress)  %>%
    compute("bigram_count")
```

Finally, we can turn to the actual method/analysis suggested by Gentzkow and Shapiro (2010). They suggest a simple chi-squared test to find the most partisan bigrams. For each bigram, we compute the corresponding chi-squared value.

```
# FIND MOST PARTISAN BIGRAMS ------------

# compute frequencies and chi-squared values
freqs <-
    bigram_count  %>%
    group_by(party, congress)  %>%
    mutate(total=sum(n), f_npl=total-n)
freqs_d <-
    filter(freqs, party=="D")  %>%
    rename(f_pld=n, f_npld=f_npl)  %>%
    select(bigram, congress, f_pld, f_npld)


## Adding missing grouping variables: `party`
```

```
freqs_r <-
    filter(freqs, party=="R") %>%
    rename(f_plr=n, f_nplr=f_npl) %>%
    select(bigram, congress, f_plr, f_nplr)
```

```
## Adding missing grouping variables: `party`
```

Based on the computed bigram frequencies, we can compute the chi-squared test as follows.

```
pol_bigrams <-
    inner_join(freqs_d, freqs_r, by=c("bigram", "congress")) %>%
    group_by(bigram, congress) %>%
    mutate(x2=((f_plr*f_npld-f_pld*f_nplr)^2)/
                ((f_plr + f_pld)*(f_plr + f_nplr)*
                    (f_pld + f_npld)*(f_nplr + f_npld))) %>%
    select(bigram, congress, x2, f_pld, f_plr) %>%
    compute("pol_bigrams")
```

15.2.5   Results: Most partisan phrases by congress

In order to present a first glimpse at the results we first select the 2,000 most partisan phrases per Congress according to the procedure above. To do so, we need to first create an index column in the corresponding Spark table.[3] We then collect the 2,000 most partisan bigrams.[4]

```
# create output data frame
output <- pol_bigrams %>%
    group_by(congress) %>%
    arrange(desc(x2)) %>%
    sdf_with_sequential_id(id="index")  %>%
    filter(index<=2000) %>%
    mutate(Party=ifelse(f_pld<f_plr, "R", "D"))%>%
    select(bigram, congress, Party, x2) %>%
```

---

[3]Recall that Spark is made to operate on distributed systems. Since the entire dataset is distributed (as subsets) across the cluster, it is not as straightforward to fetch the first N entries of a dataset as compared to a situation in which the entire dataset is stored in one data frame residing in RAM.

[4]The output of the entire pipeline is not at all large anymore at this stage; thus we can confidently call collect() to download the results (output data) from the Spark cluster into our local R memory (or the master node's R memory when working in the cloud).

```
    collect()

# disconnect from cluster
spark_disconnect(sc)
```

From the subset of the 2,000 most partisan bigrams, we then generate a table of the top 5 most partisan bigrams per Congress.

```
# packages to prepare and plot
library(data.table)
library(ggplot2)
# select top ten per congress, clean
output <- as.data.table(output)
topten <- output[order(congress, x2, decreasing = TRUE),
                 rank:=1:.N, by=list(congress)][rank %in% (1:5)]
topten[, congress:=gsub("990", "99", congress)]
topten[, congress:=gsub("980", "98", congress)]
topten[, congress:=gsub("970", "97", congress)]

# plot a visualization of the most partisan terms
ggplot(topten, mapping=aes(x=as.integer(congress), y=log(x2), color=Party)) +
    geom_text(aes(label=bigram), nudge_y = 1)+
    ylab("Partisanship score (Ln of Chisq. value)") +
    xlab("Congress") +
    scale_color_manual(values=c("D"="blue", "R"="red"), name="Party") +
    guides(color=guide_legend(title.position="top")) +
    scale_x_continuous(breaks=as.integer(unique(topten$congress))) +
    theme_minimal() +
    theme(axis.text.x = element_text(angle = 90, hjust = 1),
          axis.text.y = element_text(hjust = 1),
          panel.grid.major = element_blank(),
          panel.grid.minor = element_blank(),
          panel.background = element_blank())
```

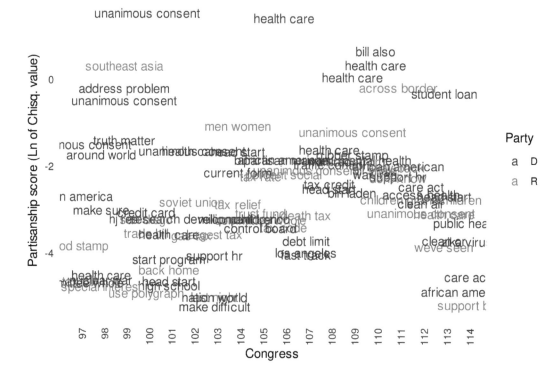

15.3 Natural Language Processing at Scale

The examples above merely scratch the surface of what is possible these days in the realm of text analysis. With increasing availability of Big Data and the recent boost in deep learning, Natural Language Processing (NLP) put forward several very powerful (and very large) language models for various text prediction tasks. Due to the improvement of these models when trained on massive amounts of text data and the rather generic application of many of these large models, it has become common practice to work directly with a pre-trained model. That is, we do not actually train the algorithm based on our own training dataset but rather build on a model that has been trained on a large text corpus and then has been made available to the public. In this section, we look at one straightforward way to build on such models with `sparklyr`.

15.3.1 Preparatory steps

Specifically, we look at a few brief examples based on the `sparknlp` package (Kincaid and Kuo, 2023) providing a `sparklyr` extension for using the John Snow Labs Spark NLP[5] library. The package can be directly installed from GitHub:

---

[5]https://www.johnsnowlabs.com/spark-nlp

`devtools::install_github("r-spark/sparknlp")` To begin, we load the corresponding packages and initialize a pre-trained NLP pipeline for sentiment analysis, which we will then apply to the congressional speeches data. Note that the `sparknlp` package needs to be loaded before we connect the R session to the Spark cluster. In the following code chunk we thus first load the package and initiate the session by connecting again to the local Spark node. In addition to loading the `sparklyr` and `dplyr` packages, we also load the `sparklyr.nested` package (Pollock, 2023). The latter is useful when working with `sparknlp`'s pipelines because the results are often returned as nested lists (in Spark table columns).

```
# load packages
library(dplyr)
library(sparklyr)
library(sparknlp)
library(sparklyr.nested)


# configuration of local spark cluster
conf <- spark_config()
conf$`sparklyr.shell.driver-memory` <- "16g"
# connect rstudio session to cluster
sc <- spark_connect(master = "local",
                    config = conf)
```

The goal of this brief example of `sparknlp` is to demonstrate how we can easily tap into very powerful pre-trained models to categorize text.To keep things simple, we return to the previous context (congressional speeches) and reload the speeches dataset. To make the following chunks of code run smoothly and relatively fast on a local Spark installation (for test purposes), we use `sample_n()` for a random draw of 10,000 speeches.

```
# LOAD ------------------


# load speeches
INPUT_PATH_SPEECHES <- "data/text/speeches/"
speeches <-
    spark_read_csv(sc,
                   name = "speeches",
                   path = INPUT_PATH_SPEECHES,
                   delimiter = "|",
                   overwrite = TRUE) %>%
```

```
      sample_n(10000, replace = FALSE)   %>%
      compute("speeches")
```

15.3.2 Sentiment annotation

In this short tutorial, we'll examine the tone (sentiment) of the congressional speeches. Sentiment analysis is a fairly common task in NLP, but it is frequently a computationally demanding task with numerous preparatory steps. sparknlp provides a straightforward interface for creating the necessary NLP pipeline in R and massively scaling the analysis on Spark. Let's begin by loading the pretrained NLP pipeline for sentiment analysis provided in sparknlp.

```
# load the nlp pipeline for sentiment analysis
pipeline <- nlp_pretrained_pipeline(sc, "analyze_sentiment", "en")
```

We can easily feed in the entire speech corpus via the target argument and point to the column containing the raw text (here "speech"). The code below divides the text into sentences and tokens (words) and returns the sentiment annotation for each sentence.

```
speeches_a <-
    nlp_annotate(pipeline,
                 target = speeches,
                 column = "speech")
```

The sentiment of the sentences is then extracted for each corresponding speech ID and coded with two additional indicator variables, indicating whether a sentence was classified as positive or negative.

```
# extract sentiment coding per speech
sentiments <-
    speeches_a %>%
    sdf_select(speech_id, sentiments=sentiment.result) %>%
    sdf_explode(sentiments)   %>%
    mutate(pos = as.integer(sentiments=="positive"),
           neg = as.integer(sentiments=="negative"))   %>%
    select(speech_id, pos, neg)
```

15.4 Aggregation and visualization

Finally, we compute the proportion of sentences with a positive sentiment per speech and export the aggregate sentiment analysis result to the R environment for further processing.[6]

```
# aggregate and download to R environment -----
sentiments_aggr <-
    sentiments %>%
    select(speech_id, pos, neg) %>%
    group_by(speech_id) %>%
    mutate(rel_pos = sum(pos)/(sum(pos) + sum(neg))) %>%
    filter(0<rel_pos) %>%
    select(speech_id, rel_pos) %>%
    sdf_distinct(name = "sentiments_aggr") %>%
    collect()


# disconnect from cluster
spark_disconnect(sc)
```

We can easily plot the aggregate speech sentiment over time because the speech ID is based on the Congress number and the sequential number of speeches in this Congress. This allows us to compare (in the simple setup of this tutorial) the sentiment of congressional speeches over time.

```
# clean
library(data.table)
sa <- as.data.table(sentiments_aggr)
sa[, congress:=substr(speech_id, 1,3)]
sa[, congress:=gsub("990", "99", congress)]
sa[, congress:=gsub("980", "98", congress)]
sa[, congress:=gsub("970", "97", congress)]

# visualize results
library(ggplot2)
ggplot(sa, aes(x=as.integer(congress),
```

---

[6]Note that even if we run this NLP pipeline for the entire text dataset of congressional speeches, the final aggregate output will easily fit into the RAM of a standard PC.

```
                y=rel_pos,
                group=congress)) +
    geom_boxplot() +
    ylab("Share of sentences with positive tone") +
    xlab("Congress") +
    theme_minimal()
```

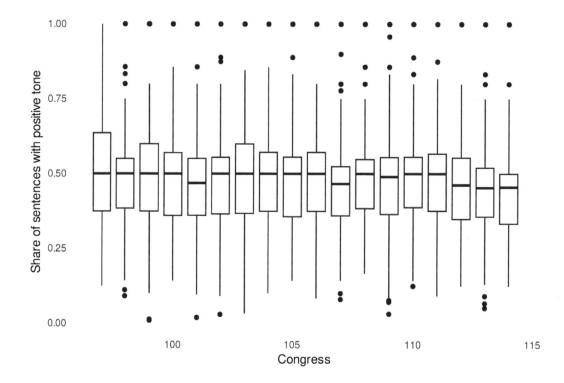

15.5   `sparklyr` and lazy evaluation

When running the code examples above, you may have noticed that the execution times vary significantly between the different code chunks, and maybe not always in the expected way. When using Apache Spark via the `sparklyr/dplyr`-interface as we did above, the evaluation of the code is intentionally (very) lazy. That is, unless a line of code really requires data to be processed (for example, due to printing the results to the console or explicitly due to calling `collect()`), Spark will not be triggered to run the actual processing of the entire data involved. When working with extremely large datasets, it makes sense to modify one's workflow to accommodate this behavior. A reasonable workflow would then be to write down the pipeline so that the heavy load processing happens at the very end (which can then take several minutes, but you will have time for other things to do.)

The following short example taken from the script developed above illustrates this point. The arguably computationally most intensive part of the previous section was the sentiment annotation via nlp_annotate():

```
system.time(
speeches_a <-
    nlp_annotate(pipeline,
                   target = speeches,
                   column = "speech")
)
```

```
##     user  system elapsed
##    0.074   0.011   0.225
```

Remember that the pre-trained pipeline used in this example includes many steps, such as breaking down speeches into sentences and words, cleaning the text, and predicting the sentiment of each sentence. When you run the code above, you'll notice that this was not the most time-consuming part to compute. That chunk of code runs in less than a second on my machine (with a local Spark node). Because we do not request the sentiment analysis results at this point, the pipeline is not actually run. It is only executed when we request it, for example, by adding the compute() call at the end.

```
system.time(
speeches_a <-
    nlp_annotate(pipeline,
                   target = speeches,
                   column = "speech") %>%
    compute(name= "speeches_a")
)
```

```
##     user  system elapsed
##    0.370   0.176  33.720
```

As you can see, this takes an order of magnitude longer, which makes perfect sense given that the pipeline is now running for the entire dataset fed into it. Unless you require the intermediate results (for example, for inspection), it thus makes sense to only process the big workload at the end of your sparklyr-analytics script.

Part V

Appendices

Appendix A: GitHub

GitHub can be a very useful platform to arrange, store, and share the code of your analytics projects even if it is typically used for collaborative software development. If you are unfamiliar with Git or GitHub, the steps below will assist you in getting started.

Initiate a new repository

1. Log in to your GitHub account and click on the plus sign in the upper right corner. From the drop-down menu select New repository.
2. Give your repository a name, for example, bigdatastat. Then, click on the big green button, Create repository. You have just created a new repository.
3. Open Rstudio, and and navigate to a place on your hard-disk where you want to have the local copy of your repository.
4. Then create the local repository as suggested by GitHub (see the page shown right after you have clicked on Create repository: "...or create a new repository on the command line"). In order to do so, you have to switch to the Terminal window in RStudio and type (or copy and paste) the commands as given by GitHub. This should look similar to the following code chunk:

```
echo "# bigdatastat" >> README.md
git init
git add README.md
git commit -m "first commit"
git remote add origin \
https://github.com/YOUR-GITHUB-ACCOUNTNAME/bigdatastat.git
git push -u origin master
```

Remember to replace YOUR-GITHUB-ACCOUNTNAME with your GitHub account name, before running the code above.

5. Refresh the page of your newly created GitHub repository. You should now see the result of your first commit.
6. Open README.md in RStudio, and add a few words describing what this repository is all about.

Clone this book's repository

1. In RStudio, navigate to a folder on your hard-disk where you want to have a local copy of this book's GitHub repository.
2. Open a new browser window, and go to https://github.com/umatter/Big Data.
3. Click on Clone or download and copy the link.
4. In RStudio, switch to the Terminal, and type the following command (pasting the copied link).

```
git clone https://github.com/umatter/BigData.git
```

You now have a local copy of the repository which is linked to the one on GitHub. You can see this by changing to the newly created directory, containing the local copy of the repository:

```
cd BigData
```

Whenever there are some updates to the book's repository on GitHub, you can update your local copy with:

```
git pull
```

(Make sure you are in the BigData folder when running git pull.)

Fork this book's repository

1. Go to https://github.com/umatter/BigData, and click on the 'Fork' button in the upper-right corner (follow the instructions).

2. Clone the forked repository (see the cloning of a repository above for details). Assuming you called your forked repository `BigData-forked`, you run the following command in the terminal (replacing `<yourgithubusername>`):

```
git clone https://github.com/`<yourgithubusername>`/BigData-forked.git
```

3. Switch into the newly created directory:

```
cd BigData-forked
```

4. Set a remote connection to the *original* repository:

```
git remote add upstream https://github.com/umatter/BigData.git
```

You can verify the remotes of your local clone of your forked repository as follows:

```
git remote -v
```

You should see something like

```
origin   https://github.com/<yourgithubusername>/BigData-forked.git (fetch)
origin   https://github.com/<yourgithubusername>/BigData-forked.git (push)
upstream     https://github.com/umatter/BigData.git (fetch)
upstream     https://github.com/umatter/BigData.git (push)
```

5. Fetch changes from the original repository. New material has been added to the original book repository, and you want to merge it with your forked repository. In order to do so, you first fetch the changes from the original repository:

```
git fetch upstream
```

6. Make sure you are on the master branch of your local repository:

```
git checkout master
```

7. Merge the changes fetched from the original repo with the master of your (local clone of the) forked repository:

```
git merge upstream/master
```

8. Push the changes to your forked repository on GitHub:

```
git push
```

Now your forked repo on GitHub also contains the commits (changes) in the original repository. If you make changes to the files in your forked repo, you can add, commit, and push them as in any repository. Example: open README.md in a text editor (e.g. RStudio), add # HELLO WORLD to the last line of README.md, and save the changes. Then:

```
git add README.md
git commit -m "hello world"
git push
```

Appendix B: R Basics

This appendix provides an overview of various key R properties, including data types and data structures.

Data types and memory/storage

Data loaded into RAM can be interpreted differently by R depending on the data *type*. Some operators or functions in R only accept data of a specific type as arguments. For example, we can store the numeric values 1.5 and 3 in the variables a and b, respectively.

```
a <- 1.5
b <- 3
a + b
```

```
## [1] 4.5
```

R interprets this data as type `double` (class 'numeric'):

```
typeof(a)
```

```
## [1] "double"
```

```
class(a)
```

```
## [1] "numeric"
```

```
object.size(a)
```

```
## 56 bytes
```

If, however, we define a and b as follows, R will interpret the values stored in a and b as text (`character`).

```
a <- "1.5"
b <- "3"
a + b
```

```
typeof(a)
```

```
## [1] "double"
```

```
class(a)
```

```
## [1] "numeric"
```

```
object.size(a)
```

```
## 56 bytes
```

Note that the symbols 1.5 take up more or less memory depending on the data-type they are stored in. This directly links to how data/information is stored/represented in binary code, which in turn is reflected in how much memory is used to store these symbols in an object as well as what we can do with it.

Example: Data types and information storage

Given the fact that computers only understand 0s and 1s, different approaches are taken to map these digital values to other symbols or images (text, decimal numbers, pictures, etc.) that we humans can more easily make sense of. Regarding text and numbers, these mappings involve *character encodings* (in which combinations of 0s and 1s represent a character in a specific alphabet) and *data types*.

Let's illustrate the main concepts with the simple numerical example from above. When we see the decimal number 139 written somewhere, we know that it means 'one-hundred-and-thirty-nine'. The fact that our computer is able to print 139 on the screen means that our computer can somehow map a sequence of 0s and 1s to the symbols 1, 3, and 9. Depending on what we want to do with the data value 139 on our computer, there are different ways of how the computer can represent this value internally. Inter alia, we could load it into RAM as a *string* ('text'/'character') or as an *integer* ('natural number') or *double* (numeric, floating point number). All of them can be printed on screen but only the latter two can be used for arithmetic computations. This concept can easily be illustrated in R.

We initiate a new variable with the value 139. By using this syntax, R by default initiates the variable as an object of type double. We then can use this variable in arithmetic operations.

```
my_number <- 139
# check the class
typeof(my_number)
```

```
## [1] "double"
```

```
# arithmetic
my_number*2
```

```
## [1] 278
```

When we change the *data type* to 'character' (string) such operations are not possible.

```
# change and check type/class
my_number_string <- as.character(my_number)
typeof(my_number_string)
```

```
## [1] "character"
```

```
# try to multiply
my_number_string*2
```

```
## Error in my_number_string * 2: non-numeric argument to binary operator
```

If we change the variable to type integer, we can still use math operators.

```
# change and check type/class
my_number_int <- as.integer(my_number)
typeof(my_number_int)
```

```
## [1] "integer"
```

```
# arithmetics
my_number_int*2
```

```
## [1] 278
```

Having all variables in the correct type is important for data analytics with various sample sizes. However, because different data types must be represented differently internally, different types may take up more or less memory, affecting performance when dealing with massive amounts of data.

We can illustrate this point with `object.size()`:

```
object.size("139")
```

```
## 112 bytes
```

```
object.size(139)
```

```
## 56 bytes
```

Data structures

For the time being, we have only looked at individual bytes of data. A single dataset can contain gigabytes of data and both text and numeric values. R has several classes of objects that provide different data structures. The data types and data structures used to store data can both affect how much memory is required to hold a dataset in RAM.

Vectors vs. Factors in R

Vectors are collections of values of the same type. They can contain either all numeric values or all character values.

For example, we can initiate a character vector containing information on the hometowns of persons participating in a survey.

```
hometown <- c("St.Gallen", "Basel", "St.Gallen")
hometown
```

```
## [1] "St.Gallen" "Basel"     "St.Gallen"
```

```
object.size(hometown)
```

```
## 200 bytes
```

Unlike in the data types example above, storing these values as type numeric to save memory is unlikely to be practical. R would be unable to convert these strings into floating point numbers. Alternatively, we could consider a correspondence table in which each unique town name in the dataset is assigned a numeric (id) code. We would save memory this way, but it would require more effort to work with the data. Fortunately, the data structure 'factor' in basic R already implements this idea in a user-friendly manner.

Factors are sets of categories. Thus, the values are drawn from a fixed set of possible values.

Considering the same example as above, we can store the same information in an object of type class factor.

```
hometown_f <- factor(c("St.Gallen", "Basel", "St.Gallen"))
hometown_f
```

```
## [1] St.Gallen Basel     St.Gallen
## Levels: Basel St.Gallen
```

```
object.size(hometown_f)
```

```
## 584 bytes
```

At first glance, the fact that hometown f consumes more memory than its character vector sibling appears strange. But we've seen this kind of 'paradox' before. Once again, the more sophisticated approach has an overhead (here not in terms of computing time but in terms of structure encoded in an object). hometown_f has more structure (i.e., a number-to-'factor level'/category label mapping). This additional structure is also data that must be saved somewhere. This disadvantage, as in previous examples of overhead costs, diminishes with larger datasets:

```
# create a large character vector
hometown_large <- rep(hometown, times = 1000)
# and the same content as factor
hometown_large_f <- factor(hometown_large)
# compare size
object.size(hometown_large)
```

```
## 24168 bytes
```

```
object.size(hometown_large_f)
```

```
## 12568 bytes
```

Matrices/Arrays

Matrices are two-dimensional collections of values of the same type, arrays are higher-dimensional collections of values of the same type.

For example, we can initiate a three-row/two-column numeric matrix as follows.

```
my_matrix <- matrix(c(1,2,3,4,5,6), nrow = 3)
my_matrix
```

```
##      [,1] [,2]
## [1,]    1    4
## [2,]    2    5
## [3,]    3    6
```

And a three-dimensional numeric array as follows.

```
my_array <- array(c(1,2,3,4,5,6), dim = 3)
my_array
```

```
## [1] 1 2 3
```

Data frames, tibbles, and data tables

Remember that in R, data frames are the most common way to represent a (table-like) dataset. Each column can contain a vector of a specific data type (or a factor), but all columns must be the same length. In the context of data analysis, each row of a data frame contains an observation, and each column contains a characteristic of that observation.

The previous implementation of data frames in R made it difficult to work with large datasets.[7] Several newer R implementations of the data-frame concept were introduced with the aim to speed up data processing. One is known as tibble, and it is implemented and used in the tidyverse packages. The other is known as data table, and it is implemented in the data table-package. Most of the shortcomings of the original 'data.frame' implementation, however, have been addressed in

---

[7]This was not an issue in the early days of R because datasets that were rather large by today's standards (in the Gigabytes) could not have been handled properly by normal computers anyway (due to a lack of RAM).

subsequent R versions, making traditional `data.frames`, `tibbles`, and `data.tables` more similarly suitable for working with large datasets (for in-memory processing).

Here is how we define a `data.table` in R:

```r
# load package
library(data.table)
# initiate a data.table
dt <- data.table(person = c("Alice", "Ben"),
                 age = c(50, 30),
                 gender = c("f", "m"))
dt
```

```
##     person age gender
## 1:  Alice  50      f
## 2:    Ben  30      m
```

Lists

Similar to data frames and data tables, lists can contain different types of data in each element. For example, a list could contain several other lists, data frames, and vectors with differing numbers of elements.

This flexibility can easily be demonstrated by combining some of the data structures created in the examples above:

```r
my_list <- list(my_array, my_matrix, dt)
my_list
```

```
## [[1]]
## [1] 1 2 3
##
## [[2]]
##      [,1] [,2]
## [1,]    1    4
## [2,]    2    5
## [3,]    3    6
##
## [[3]]
##     person age gender
## 1:  Alice  50      f
## 2:    Ben  30      m
```

R-tools to investigate structures and types

package	function	purpose
utils	str()	Compactly display the structure of an arbitrary R object.
base	class()	Prints the class(es) of an R object.
base	typeof()	Determines the (R-internal) type or storage mode of an object.

Appendix C: Install Hadoop

You might wish to install Hadoop locally on your computer in order to perform the Hadoop example in Chapter 6. The next few stages assist you in configuring everything. Please be aware that between the time I wrote this book and the time you read it, Hadoop may have undergone some changes. Consult https://hadoop .apache.org/ for further details on releases and to install the most recent version. However, the instructions for installing Hadoop in the following should be nearly comparable. Please take note that the steps below assume you are using *Ubuntu Linux*. See the README file in https://github.com/umatter/bigdata for additional hints regarding the installation of software used in this book.

```
# download binary
wget https://dlcdn.apache.org/hadoop/common/hadoop-2.10.1/hadoop-2.10.1.tar.gz
# download checksum
wget \
https://dlcdn.apache.org/hadoop/common/hadoop-2.10.1/hadoop-2.10.1.tar.gz.sha512


# run the verification
shasum -a 512 hadoop-2.10.1.tar.gz
# compare with value in mds file
cat hadoop-2.10.1.tar.gz.sha512


# if all is fine, unpack
tar -xzvf hadoop-2.10.1.tar.gz
# move to proper place
sudo mv hadoop-2.10.1 /usr/local/hadoop



# then point to this version from hadoop
# open the file /usr/local/hadoop/etc/hadoop/hadoop-env.sh



# in a text editor and add (where export JAVA_HOME=...)
export JAVA_HOME=$(readlink -f /usr/bin/java | sed "s:bin/java::")
```

```
# clean up
rm hadoop-2.10.1.tar.gz
rm hadoop-2.10.1.tar.gz.sha512
```

After running all of the steps above, run the following line in the terminal to check the installation

```
# check installation
/usr/local/hadoop/bin/hadoop
```

Part VI

Bibliography and Index

Bibliography

Abadi, M., Agarwal, A., Barham, P., Brevdo, E., Chen, Z., Citro, C., Corrado, G. S., Davis, A., Dean, J., Devin, M., Ghemawat, S., Goodfellow, I., Harp, A., Irving, G., Isard, M., Jia, Y., Jozefowicz, R., Kaiser, L., Kudlur, M., Levenberg, J., Mané, D., Monga, R., Moore, S., Murray, D., Olah, C., Schuster, M., Shlens, J., Steiner, B., Sutskever, I., Talwar, K., Tucker, P., Vanhoucke, V., Vasudevan, V., Viégas, F., Vinyals, O., Warden, P., Wattenberg, M., Wicke, M., Yu, Y., and Zheng, X. (2015). TensorFlow: Large-scale machine learning on heterogeneous systems. Software available from tensorflow.org.

Adler, D., Gläser, C., Nenadic, O., Oehlschlägel, J., Schuemie, M., and Zucchini, W. (2022). *ff: Memory-Efficient Storage of Large Data on Disk and Fast Access Functions*. R package version 4.0.7.

Aldrich, E. M. (2014). Chapter 10 - GPU Computing in Economics. In Schmedders, K. and Judd, K. L., editors, *Handbook of Computational Economics Vol. 3*, volume 3 of *Handbook of Computational Economics*, pages 557–598. Elsevier.

Aldrich, E. M., Fernandez-Villaverde, J., Ronald Gallant, A., and Rubio-Ramirez, J. F. (2011). Tapping the supercomputer under your desk: Solving dynamic equilibrium models with graphics processors. *Journal of Economic Dynamics and Control*, 35(3):386–393.

Allaire, J. and Chollet, F. (2022). *keras: R Interface to 'Keras'*. R package version 2.11.0.

Angrist, J. D. and Pischke, J.-S. (2008). *Mostly Harmless Econometrics: An Empiricist's Companion*. Princeton University Press.

Bates, D., Maechler, M., and Jagan, M. (2022). *Matrix: Sparse and Dense Matrix Classes and Methods*. R package version 1.5-3.

Bengtsson, H. (2021). A Unifying Framework for Parallel and Distributed Processing in R using Futures. *The R Journal*, 13(2):273–291.

Burns, P. (2011). *The R Inferno*. Lulu Press, Inc.

Chang, W., Luraschi, J., and Mastny, T. (2020). *profvis: Interactive Visualizations for Profiling R Code*. R package version 0.3.7.

Cheng, H.-T., Haque, Z., Hong, L., Ispir, M., Mewald, C., Polosukhin, I., Roumpos, G., Sculley, D., Smith, J., Soergel, D., Tang, Y., Tucker, P., Wicke,

M., Xia, C., and Xie, J. (2017). TensorFlow Estimators: Managing Simplicity vs. Flexibility in High-Level Machine Learning Frameworks. In *Proceedings of the 23rd ACM SIGKDD International Conference on Knowledge Discovery and Data Mining*, KDD '17, page 1763–1771, New York, NY, USA. Association for Computing Machinery.

Chollet, F. et al. (2015). Keras. https://keras.io [last visited: 16/05/2023].

Cohen, L. and Malloy, C. J. (2014). Friends in high places. *American Economic Journal: Economic Policy*, 6(3):63–91.

de Jonge, E., Wijffels, J., and van der Laan, J. (2023). *ffbase: Basic Statistical Functions for Package 'ff'*. R package version 0.13.2.

Determan, C. (2019). *gpuR: GPU Functions for R Objects*. R package version 2.0.3.

Dhillon, P., Lu, Y., Foster, D. P., and Ungar, L. (2013). New subsampling algorithms for fast least squares regression. In *Advances in Neural Information Processing Systems 26*, pages 360–368.

Donoho, D. (2017). 50 years of data science. *Journal of Computational and Graphical Statistics*, 26(4):745–766.

Dowle, M. and Srinivasan, A. (2022). *data.table: Extension of 'data.frame'*. R package version 1.14.6.

Fatahalian, K., Sugerman, J., and Hanrahan, P. (2004). Understanding the Efficiency of GPU Algorithms for Matrix-Matrix Multiplication. In *Proceedings of the ACM SIGGRAPH/EUROGRAPHICS Conference on Graphics Hardware*, HWWS '04, page 133–137, New York, NY, USA. Association for Computing Machinery.

Frick, H., Chow, F., Kuhn, M., Mahoney, M., Silge, J., and Wickham, H. (2022). *rsample: General Resampling Infrastructure*. R package version 1.1.1.

Fultz, N. and Daróczi, G. (2019). *AWR.Athena: 'AWS' Athena 'DBI' Wrapper*. R package version 2.0.7-0.

Gaure, S. (2013a). lfe: Linear Group Fixed Effects. *The R Journal*, 5(2):104–116.

Gaure, S. (2013b). OLS with multiple high dimensional category variables. *Computational Statistics & Data Analysis*, 66:8–18.

Gentzkow, M. and Shapiro, J. M. (2010). What drives media slant? Evidence from U.S. daily newspapers. *Econometrica*, 78(1):35–71.

Gentzkow, M., Shapiro, J. M., and Taddy, M. (2019). Measuring group differences in high-dimensional choices: Method and application to congressional speech. *Econometrica*, 87(4):1307–1340.

Hester, J. and Vaughan, D. (2021). *bench: High Precision Timing of R Expressions*. R package version 1.1.2.

Hester, J., Wickham, H., and Csárdi, G. (2023). *fs: Cross-Platform File System Operations Based on 'libuv'*. R package version 1.6.0.

Højsgaard, S. and Halekoh, U. (2023). *doBy: Groupwise Statistics, LSmeans, Linear Estimates, Utilities*. R package version 4.6.16.

Johnston, M. and Robinson, D. (2022). *gutenbergr: Download and Process Public Domain Works from Project Gutenberg*. R package version 0.2.3.

Kane, M. J., Emerson, J., and Weston, S. (2013). Scalable strategies for computing with massive data. *Journal of Statistical Software*, 55(14):1–19.

Karami, A., Gangopadhyay, A., Zhou, B., and Kharrazi, H. (2017). Fuzzy approach topic discovery in health and medical corpora. *International Journal of Fuzzy Systems*, 20:1334–1345.

Kincaid, D. and Kuo, K. (2023). *sparknlp: R Interface to John Snow Labs Spark NLP*. R package version 0.16.0.

Kratochvil, M. (2022). *scattermore: Scatterplots with More Points*. R package version 0.8.

Kratochvíl, M., Bednárek, D., Sieger, T., Fišer, K., and Vondrášek, J. (2020). ShinySOM: graphical SOM-based analysis of single-cell cytometry data. *Bioinformatics*, 36(10):3288–3289.

Kuhn, M. and Vaughan, D. (2022). *parsnip: A Common API to Modeling and Analysis Functions*. R package version 1.0.3.

Kuhn, M., Vaughan, D., and Hvitfeldt, E. (2022). *yardstick: Tidy Characterizations of Model Performance*. R package version 1.1.0.

Kuhn, M. and Wickham, H. (2020). Tidymodels: a collection of packages for modeling and machine learning using tidyverse principles. https://www.tidymodels.org [last visited: 16/05/2023].

Leeper, T. J. (2020). *aws.s3: AWS S3 Client Package*. R package version 0.3.21.

Luraschi, J., Kuo, K., Ushey, K., Allaire, J., Falaki, H., Wang, L., Zhang, A., Li, Y., Ruiz, E., and The Apache Software Foundation (2022). *sparklyr: R Interface to Apache Spark*. R package version 1.7.9.

Matloff, N. (2015). *Parallel Computing for Data Science*. CRC Press, Boca Raton, FL.

Mersmann, O. (2021). *microbenchmark: Accurate Timing Functions*. R package version 1.4.9.

Metamarkets Group Inc. (2023). *RDruid: Druid Connector for R*. R package version 0.2.3.

Meyer, D., Zeileis, A., and Hornik, K. (2006). The strucplot framework: Visualizing multi-way contingency tables with vcd. *Journal of Statistical Software*, 17(3):1–48.

Meyer, D., Zeileis, A., and Hornik, K. (2023). *vcd: Visualizing Categorical Data*. R package version 1.4-11.

Microsoft and Weston, S. (2022). *foreach: Provides Foreach Looping Construct*. R package version 1.5.2.

Microsoft Corporation and Weston, S. (2022). *doSNOW: Foreach Parallel Adaptor for the 'snow' Package*. R package version 1.0.20.

Müller, K., Wickham, H., James, D. A., and Falcon, S. (2022). *RSQLite: SQLite Interface for R*. R package version 2.2.20.

Pollock, M. (2023). *sparklyr.nested: A 'sparklyr' Extension for Nested Data*. R package version 0.0.3.

R Core Team (2021). *R: A Language and Environment for Statistical Computing*. R Foundation for Statistical Computing, Vienna, Austria.

R Core Team (2022). *foreign: Read Data Stored by 'Minitab', 'S', 'SAS', 'SPSS', 'Stata', 'Systat', 'Weka', 'dBase', ...* R package version 0.8-82.

R Special Interest Group on Databases (R-SIG-DB), Wickham, H., and Müller, K. (2022). *DBI: R Database Interface*. R package version 1.1.3.

Richardson, N., Cook, I., Crane, N., Dunnington, D., François, R., Keane, J., Moldovan-Grünfeld, D., Ooms, J., and Apache Arrow (2022). *arrow: Integration to 'Apache' 'Arrow'*. https://github.com/apache/arrow/, https://arrow.apache.org/docs/r/.

Robinson, D., Hayes, A., and Couch, S. (2022). *broom: Convert Statistical Objects into Tidy Tibbles*. R package version 1.0.2.

Shakespeare, W. (1599/2020). *Julius Caesar*. Open Road Media, New York.

Stock, J. H. and Watson, M. W. (2003). *Introduction to Econometrics*. Pearson Education.

Taddy, M. (2017). One-step estimator paths for concave regularization. *Journal of Computational and Graphical Statistics*, 26(3):525–536.

Taddy, M. (2019). *Business Data Science*. McGraw-Hill, New York.

Urbanek, S. (2022). *RJDBC: Provides Access to Databases Through the JDBC Interface*. R package version 0.2-10.

Varian, H. R. (2014). Big data: New tricks for econometrics. *Journal of Economic Perspectives*, 28(2):3–28.

Venkataraman, S., Meng, X., Cheung, F., and The Apache Software Foundation (2021). *SparkR: R Front End for Apache Spark*. R package version 3.1.2.

Walkowiak, S. (2016). *Big Data Analytics with R*. PACKT Publishing, Birmingham, UK.

Wickham, H. (2011). The split-apply-combine strategy for data analysis. *Journal of Statistical Software*, 40.

Wickham, H. (2016). *ggplot2: Elegant Graphics for Data Analysis*. Springer-Verlag New York.

Wickham, H. (2021). *pryr: Tools for Computing on the Language*. R package version 0.1.5.

Wickham, H. (2022a). *lobstr: Visualize R Data Structures with Trees*. R package version 1.1.2.

Wickham, H. (2022b). *stringr: Simple, Consistent Wrappers for Common String Operations*. R package version 1.5.0.

Wickham, H., Averick, M., Bryan, J., Chang, W., McGowan, L. D., François, R., Grolemund, G., Hayes, A., Henry, L., Hester, J., Kuhn, M., Pedersen, T. L., Miller, E., Bache, S. M., Müller, K., Ooms, J., Robinson, D., Seidel, D. P., Spinu, V., Takahashi, K., Vaughan, D., Wilke, C., Woo, K., and Yutani, H. (2019). Welcome to the tidyverse. *Journal of Open Source Software*, 4(43):1686.

Wickham, H. and Bryan, J. (2022). *bigrquery: An Interface to Google's 'BigQuery' 'API'*. R package version 1.4.1.

Wickham, H., François, R., Henry, L., Müller, K., and Vaughan, D. (2023). *dplyr: A Grammar of Data Manipulation*. R package version 1.1.0.

Wickham, H. and Grolemund, G. (2016). *R for Data Science*. O'Reilly Media, Inc.

Wickham, H., Hester, J., Chang, W., and Bryan, J. (2022). *devtools: Tools to Make Developing R Packages Easier*. R package version 2.4.5.

Wilkinson, L., Wills, D., Rope, D., Norton, A., and Dubbs, R. (2005). *The Grammar of Graphics*. Statistics and Computing. Springer New York.

Yang, F., Tschetter, E., Léauté, X., Ray, N., Merlino, G., and Ganguli, D. (2014). Druid: A Real-Time Analytical Data Store. In *Proceedings of the 2014 ACM SIGMOD International Conference on Management of Data*, SIGMOD '14, page 157–168, New York, NY, USA. Association for Computing Machinery.

Zaharia, M., Xin, R. S., Wendell, P., Das, T., Armbrust, M., Dave, A., Meng, X., Rosen, J., Venkataraman, S., Franklin, M. J., Ghodsi, A., Gonzalez, J., Shenker, S., and Stoica, I. (2016). Apache Spark: A unified engine for big data processing. *Commun. ACM*, 59(11):56–65.

Zeileis, A. and Hothorn, T. (2002). Diagnostic checking in regression relationships. *R News*, 2(3):7–10.

Zeileis, A., Meyer, D., and Hornik, K. (2007). Residual-based shadings for visualizing (conditional) independence. *Journal of Computational and Graphical Statistics*, 16(3):507–525.

Index